Ernst Schering Research Foundation Workshop 23
Excitatory Amino Acids

Springer
Berlin
Heidelberg
New York
Barcelona
Budapest
Hong Kong
London
Milan
Paris
Santa Clara
Singapore
Tokyo

Ernst Schering Research Foundation
Workshop 23

Excitatory Amino Acids

From Genes to Therapy

P. H. Seeburg, I. Bresink, L. Turski
Editors

With 45 Figures in 125 Separate Illustrations

Springer

Series Editors: G. Stock and U.-F. Habenicht

ISSN 0947-6075
ISBN 3-540-63387-1 Springer-Verlag Berlin Heidelberg New York

CIP data applied for

Die Deutsche Bibliothek – CIP-Einheitsaufnahme
Schering-Forschungsgesellschaft <Berlin>: Ernst Schering Research Foundation Workshop. - Berlin; Heidelberg; New York; Barcelona; Budapest; Hong Kong; London; Milan; Paris; Santa Clara; Singapore; Tokyo: Springer.
ISSN 0947-6075
23. Excitatory amino acids: from genes to therapy. - 1998
Excitatory amino acids: from genes to therapy / P.H. Seeburg ... ed. - Berlin; Heidelberg; New York; Barcelona; Budapest; Hong Kong; London; Milan; Paris; Santa Clara; Singapore; Tokyo: Springer, 1998
(Ernst Schering Research Foundation Workshop; 23)
ISBN 3-540-63387-1

Printed in Germany

Typesetting: TBS, Sandhausen
Printing: Druckhaus Beltz, Hemsbach. Binding: J. Schäffer GmbH & Co. KG, Grünstadt
SPIN: 10639398 13/3135–5 4 3 2 1 0 – Printed on acid-free paper

Preface

Glutamate Antagonists: Quo Vade?

Since the beginning of the 1990s, glutamate antagonists have repeatedly excited the scientific community through discoveries of their intriguing role in physiology and pathology. Until today they have remained the major hope for the therapy of neurodegenerative disorders in humans. NMDA antagonists have been designed for the treatment of acute-onset neurodegeneration such as stroke and trauma, and have recently reached the clinics. Unfortunately, the anticipated success of NMDA antagonists in patients has been limited due to unacceptable side effects.

In the meantime, AMPA antagonists have left the realm of drug research and are rapidly moving towards the clinic. The expectations and medical needs, which have changed little over time, are placing development of AMPA antagonists under public attention. The enthusiasm which surrounds AMPA antagonists is great, since they appear to be safer and to have fewer side effects than the NMDA antagonists.

Research on glutamate receptors has entered the phase of discovering molecular details of the channels and genetic principles, which determine phenotypic characteristics and sensitivity of neurons to the deleterious influences of nature.

The engineering of transgenic mice with modified expression of single-receptor subunits has enabled insights into the function, dysfunction, and possible pathology causally related to glutamate receptors.

The participants of the workshop

The meeting on "Excitatory Amino Acids: From Genes to Therapy" addresses today's emerging questions and discusses new avenues which glutamate research may take in the future. The entry of AMPA antagonists into the clinic for first evaluations in humans dominates the field of today's glutamate research. The picture of kainate receptors is slowly emerging and coming into focus. These enigmatic receptors may dominate the field soon, leading to new excitements and novel research efforts.

P. H. Seeburg
I. Bresink
L. Turski

Table of Contents

List of Editors and Contributors

Editors

P.H. Seeburg
Max Planck Institute for Medical Research, Department of Molecular Neuroscience, Jahnstr. 29, 69120 Heidelberg, Germany

I. Bresink
Research Laboratories of Schering AG, Müllerstr. 178, 13342 Berlin, Germany

L. Turski
Eisai London Research Laboratories, Bernard Katz Building, University College London, Gower Street, London WC1E 6BT, UK

Contributors

D.F. Babcock
Department of Pharmacological and Physiological Sciences, The University of Chicago, 947 East 58th Street (MC 0926), Chicago, IL 60637, USA

I. Bresink
Research Laboratories of Schering AG, Müllerstr. 178, 13342 Berlin, Germany

V.P. Bindokas
Department of Pharmacological and Physiological Sciences, The University of Chicago, 947 East 58th Street (MC 0926), Chicago, IL 60637, USA

R. Brusa
Schering Plough Research Institute, Via Olgettina 58, 20132 Milan, Italy

N. Burnashev
Max-Planck-Institute of Medical Research, Department of Cell Physiology, Jahnstr. 29, 69120 Heidelberg, Germany

H.S.-V. Chen
CNS Research Institute, Brigham and Women's Hospital and Harvard Medical School, 221 Longwood Avenue-LMRC, Boston, MA 02115, USA

Y.-B. Choi
CNS Research Institute, Brigham and Women's Hospital and Harvard Medical School, 221 Longwood Avenue-LMRC, Boston, MA 02115, USA

G.L. Collingridge
Department of Anatomy, Medical School, University of Bristol, Bristol, BS8 1TD, UK

A. Doherty
Department of Anatomy, Medical School, University of Bristol, Bristol, BS8 1TD, UK

U. Endermann
Research Laboratories of Schering AG, Müllerstr. 178, 13342 Berlin, Germany

O. Garaschuk
Institute of Physiology I, University of Saarland, 66421 Homburg, Germany

J.M. Henley
Department of Anatomy, Medical School, University of Bristol, Bristol, BS8 1TD, UK

M. Higuchi
Max Planck Institute for Medical Research, Department of Molecular Neuroscience, Jahnstr. 29, 69120 Heidelberg, Germany

A. Irving
Department of Biomedical Sciences, Institute of Medical Sciences, Forrester Hill, University of Aberdeen, Aberdeen, AB25 2ZD, UK

J.B. Jordan
Department of Pharmacological and Physiological Sciences, The University of Chicago, 947 East 58th Street (MC 0926), Chicago, IL 60637, USA

A. Konnerth
Institute of Physiology I, University of Saarland, 66421 Homburg, Germany

J.-P. Lee
Department of Pharmacological and Physiological Sciences, The University of Chicago, 947 East 58th Street (MC 0926), Chicago, IL 60637, USA

S.A. Lipton
CNS Research Institute, Brigham and Women's Hospital and Harvard Medical School, 221 Longwood Avenue-LMRC, First Floor, Boston, MA 02115, USA

R.J. Miller
Department of Pharmacological and Physiological Sciences, The University of Chicago, 947 East 58th Street (MC 0926), Chicago, IL 60637, USA

G. Müller,
Neurology Research, Warren 407, Massachusetts General Hospital, Harvard Medical School, Fruit Street, Boston, MA 02114, USA

Z. Nusser
Medical Research Council, Anatomical Neuropharmacology Unit, University of Oxford, Mansfield Road, Oxford OX1 3TH, UK

D.E. Pellegrini-Giampietro
Department of Preclinical and Clinical Pharmacology "Mario Aiazzi Mancini", University of Florence, Viale G.B. Morgagni 65, 50134 Florence, Italy

T. Plant
Institute of Physiology I, University of Saarland, 66421 Homburg, Germany

P.H. Seeburg
Max Planck Institute for Medical Research, Department of Molecular Neuroscience, Jahnstr. 29, 69120 Heidelberg, Germany

P.J. Shaw
Department of Clinical Neurology, Royal Victoria Infirmary, University of Newcastle upon Tyne, Newcastle upon Tyne NE1 4LP, UK

R. Shigemoto
Medical Research Council, Anatomical Neuropharmacology Unit, University of Oxford, Mansfield Road, Oxford OX1 3TH, UK, and Department of Morphological Brain Science, Faculty of Medicine, Kyoto University, Sakyo-ku, Kyoto 606, Japan

R. Sprengel
Max Planck Institute for Medical Research, Department of Molecular Neuroscience, Jahnstr. 29, 69120 Heidelberg, Germany

N.J. Sucher
CNS Research Institute, Brigham and Women's Hospital and Harvard Medical School, 221 Longwood Avenue-LMRC, Boston, MA 02115, USA

L. Turski
Eisai London Research Laboratories, Bernard Katz Building, University College London, Gower Street, London WC1E 6BT, UK

1 Glutamate Receptor Channels, RNA Editing, and Epilepsy

P.H. Seeburg, R. Brusa, M. Higuchi, and R. Sprengel

1.1 GluRs in Excitatory Neurotransmission

In the mammalian central nervous system (CNS) fast excitatory neurotransmission is mediated by ionotropic glutamate receptors (GluRs) of the AMPA/kainate and NMDA receptor families (Collingridge and Lester 1989; Mayer and Westbrook 1987; Watkins et al. 1990). The AMPA receptors mediate the fast component of excitatory postsynaptic currents, whereas the slow component is contributed by NMDA receptors (e.g., Stern et al. 1992). The latter can be viewed as coincidence detectors of pre- and postsynaptic activity, since the gating of the integral ion channel requires two closely simultaneous events, presynaptic release of glutamate and depolarization of the postsynaptic membrane. Depolarization is induced primarily by the activation of synaptically colocalized AMPA receptors. Coincidence detection by the NMDA receptor rests on its voltage-dependent channel block by extracellular

Mg^{2+}. NMDA receptors are designed for high Ca^{2+} permeability, and Ca^{2+} influx through the NMDA receptor channel is thought to be essential for activity-dependent synaptic modulation (reviewed in Bliss and Collingridge 1993). Furthermore, excessively high Ca^{2+} influx through NMDA receptors has pathophysiological consequences, including epileptiform activities and neurodegeneration (Choi 1988). By contrast, AMPA receptors in most neurons have low divalent to monovalent permeability ratios ($P_{Ca}/P_{monovalent}$). However, certain neuronal populations, such as GABA-ergic interneurons, express AMPA receptors that are highly Ca^{2+} permeable. This Ca^{2+} permeability is not subject to voltage-dependent Mg^{2+} block but is sensitive to intracellular polyamines (reviewed in Jonas and Burnashev 1995).

1.2 The GluR Subunits

Ionotropic GluRs are composed of subunits, and numerous subtypes differing in subunit composition and functional properties exist for each receptor type. The work of several laboratories has led to the characterization of four AMPA receptor subunits (GluR-A, -B, -C, and -D, or GluR1–GluR4), five "high-affinity" kainate receptor subunits (KA-1, KA-2, GluR5, GluR6, and GluR7), and five NMDA receptor subunits (NR1, NR2 A, NR2B, NR2C, and NR2D) (Hollmann and Heinemann 1994; Sprengel and Seeburg 1994). Each subunit is a glycosylated membrane-inserted polypeptide with an average length of 900 amino acids. The NR2 subunits of the NMDA receptors are larger (1300 amino-acid residues), featuring extended intracellular carboxy-termini (Hollmann and Heinemann 1994; Monyer et al. 1992). All subunits are predicted to contain four hydrophobic segments (M1–M4) which determine the transmembrane topology of these polypeptides. Experimental evidence indicates that GluR subunits have the amino-terminus on the outside of the cell and carboxy-terminus on the inside, and that M2 forms a pore loop structure entering and exiting the cell membrane from the intracellular face (Wo and Oswald 1994; Hollmann et al. 1994; Kuner et al. 1996). Amino acids within M2 line the inner channel pore and particular amino acids in this segment determine the distinct ion selectivities of the AMPA and NMDA receptor channels (Hollmann and Heinemann 1994).

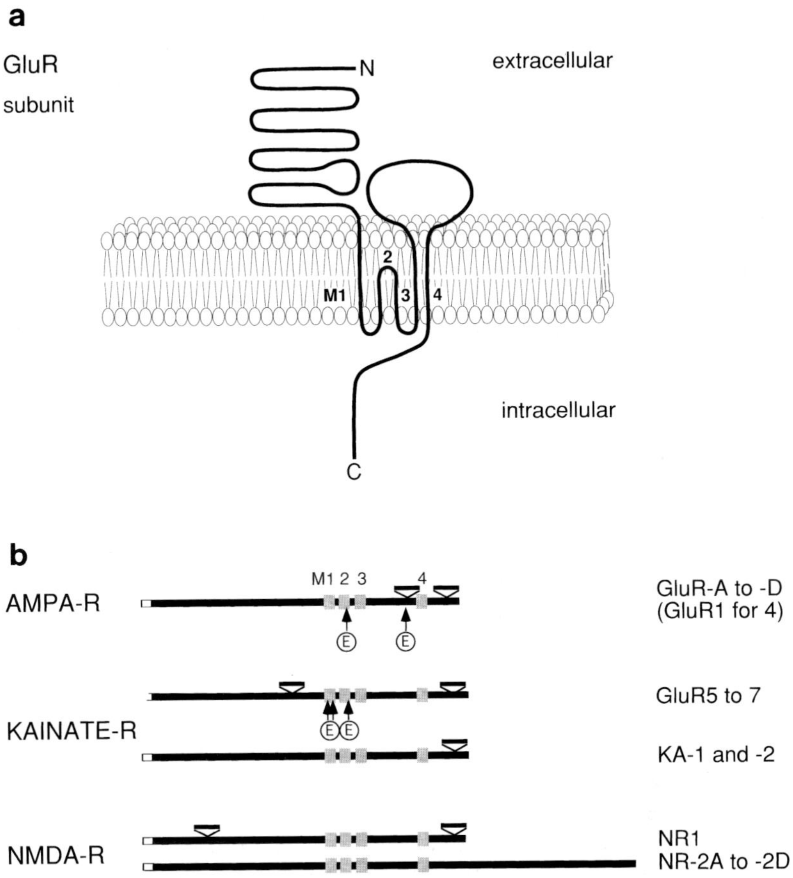

Fig. 1a,b. Schematic representation of ionotropic glutamate receptor subunits. Subunits for NMDA, AMPA, and kainate receptor subtypes of glutamate receptors are about 900 amino acids in length. The NR2 subunits of NMDA receptors are larger (1300 residues) due to extended intracellular carboxy-termini. All subunits feature an amino-terminal signal peptide (*white box*) and four hydrophobic segments (M1–M4, *gray boxes*). *Arrows*, positions for RNA editing in AMPA receptor subunits GluR-B, -C, and -D and kainate receptor subunits GluR5 and GluR6. Alternative splice products occur within the amino-terminal region, in the M3 to M4 loop, and in the hydrophilic carboxy-terminal part of the subunits (*small vertical bars*)

Alternative splicing produces subunit variants with different carboxy-termini or additional or substituted amino-acid sequences in the amino-terminal, in the extracellular region, or in the segment between M3 and M4 (Fig. 1) (Hollmann and Heinemann 1994). In addition, in AMPA and "high-affinity" kainate receptor subunits, recoding through RNA editing, a mechanism for site-selective adenosine deamination in pre-mRNA affects several amino acid positions with profound functional consequences (reviewed in Seeburg 1996). The physiologically most important one appears to be an exchange of glutamine (Q) for arginine (R) within the pore lining M2 segment of the AMPA receptor subunit GluR-B. In heteromeric AMPA receptors the GluR-B subunit dominates the channel's divalent cation conductance, rendering the channel impermeable to Ca^{2+} ions due to the unique arginine residue in M2, introduced by posttranscriptional RNA editing (Sommer et al. 1991).

1.3 Distinct AMPA Receptor Subtypes in CNS

AMPA receptors in mammalian central neurons differ considerably with respect to gating kinetics and $P_{Ca}/P_{monovalent}$ permeability ratios (reviewed in Jonas and Burnashev 1995). GABA-ergic interneurons, auditory relay neurons, septal neurons, and dorsal horn spinal neurons express AMPA receptors with faster gating kinetics and higher Ca^{2+} permeability than principal neurons. Attempts to establish a possible correlation of these functional differences with the subunit composition of AMPA receptors at the single cell level document that neurons control the level of Ca^{2+} permeability of their AMPA receptors by the extent of GluR-B subunit expression relative to the expression of the other AMPA receptor subunit genes (Geiger et al. 1995). Hence, while interneurons express, for example, high levels of GluR-A or GluR-D, but little GluR-B, principal neurons in which AMPA receptor channels exhibit low permeability to Ca^{2+} ions express high GluR-B to GluR-A, -C and -D ratios. In the absence of constraints for subunit assembly, if GluR-B expression in a neuron constitutes 10% of other AMPA receptor subunits, 40% of the channels are predicted to contain at least one GluR-B subunit. The presence of a single GluR-B subunit shuts down the Ca^{2+} permeability of heteromeric AMPA receptors. As outlined above, this functional dominance of GluR-B derives from the arginine residue in its M2 segment, which is created by RNA editing. The positive charge in the channel wall, contributed by GluR-B, further renders the heteromeric, GluR-B containing AMPA receptor insensitive to the voltage-dependent block by intracellular polyamines. This block affects AMPA receptors without GluR-B and generates a strong double rectification in the current voltage relation of the glutamate-gated currents (Jonas and Burnashev 1995).

1.4 RNA Editing in GluR Pre-mRNAs

The coding potential of a primary transcript can be altered by RNA editing, which in different plant and animal species can operate by different molecular mechanisms (Scott 1995), and which appears to be an ancient evolutionary strategy for gene diversification (Simpson and Emeson 1996). It came as a surprise that ion channels critical for neurotransmission in the mammalian brain are subject to RNA editing (Seeburg 1996). We have identified five genes in the ligand-gated GluR family, the primary transcripts of which undergo adenosine selective RNA editing in a total of eight positions (Fig. 1). GluR transcript editing targets exclusively gene-specified adenosines that appear as guanosines in cloned cDNA. For each of these sites (see below), the transmitter-activated channel acquires altered biophysical properties.

Recombinant expression of wild-type and mutated AMPA receptors identified a critical residue in the putative channel-lining segment M2, which controls Ca^{2+} ion permeability (Burnashev et al. 1992a; Hollmann et al. 1991). The arginine (R) in the M2 of the GluR-B subunit renders AMPA channels impermeable to Ca^{2+}, whereas GluR-A, -C, and -D subunits which contain glutamine (Q) in the homologous position assemble into channels with high Ca^{2+} permeability Burnashev et al. 1992a,b; Hollmann et al. 1991). Moreover, AMPA receptors assembled without GluR-B are subject to a voltage-dependent block by intracellular polyamines and thus exhibit a doubly-rectifying current-voltage (I–V) relationship (Jonas and Burnashev 1995). The genomic DNA encoding the M2 channel segment of GluR-B specifies a CAG codon (encoding glutamine) for the M2 Q/R position while all cDNA of GluR-B carries a CGG (arginine) codon for the same channel position.

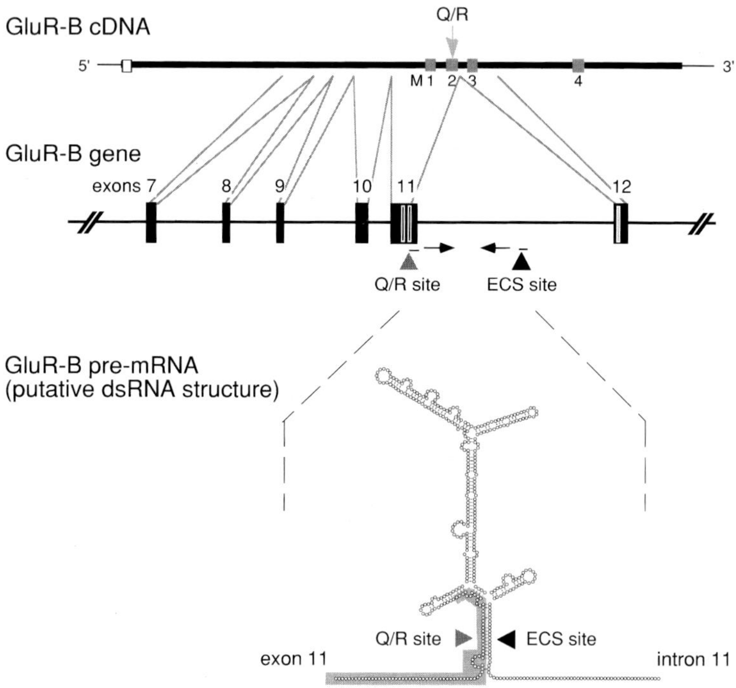

Fig. 2. *Above,* Q/R site editing of the GluR-B pre-mRNA. Linear representations of GIuR-B cDNA and the corresponding gene segment of the murine GIuR-B gene. In the cDNA: *black bar*, coding region; *gray boxes*, hydrophobic segments M1–M4; *hatched box*, "flip/flop." *Arrow*, the relative position for the Q/R site on M2. *Black boxes*, exonic sequences of the GluR-B gene; *gray box*, membrane-spanning segments (*M*). *Horizontal lines between the exons*, intronic sequences; *hatched lines projecting from cDNA to gene*, the cDNA portions contributed by exons 7–12. *Below*, the putative secondary structure which is formed by GluR-B precursor RNA sequences of exon 11 and intron 11. *Small open circles*, nucleotides; the position of the edited nucleotide of the Q/R-site and of the ECS are indicated by *black and gray arrowheads*, respectively

Multiple genes and alternative exons have been excluded as a source for the arginine codon in GluR-B mRNA, and RNA editing has been postulated as the mechanism for the conversion of the CAG to the CGG codon. Upon comparison of gene to cDNA, CAG to CGG conversions were found at homologous positions in the M2 segment of the high-affinity kainate receptor subunits GluR5 and GluR6 (Sommer et al. 1991). There is no evidence for editing in the GluR7 transcript. When brain RNA of adult rats was analyzed by reverse-transcription polymerase chain reaction, the GluR-B (R/Q) site was found to be edited almost completely (99% of the analyzed polymerase chain reaction products had an CGG arginine codon) whereas the Q/R site in GluR5 and GluR6 is edited to lower extents (40% and 80%, respectively; Bernard and Khrestchatisky 1994). GluR editing occurs in the nucleus and precedes splicing, since unspliced GluR transcripts are edited to a high level (Higuchi et al. 1993).

Transfection of GluR-B minigenes into PC12 cells has documented that the proximal part of the intron 3' to the unedited exonic site is required for Q/R site editing. This intron portion contains an imperfect inverted repeat preceding a ten-nucleotide (editing-site complementary) sequence (ECS) with perfect complementarity to the exon sequence centered on the unedited CAG codon (Fig. 2). Single nucleotide substitutions in this short intronic sequence, or in the exonic complement, severely diminished or even abrogated Q/R site editing which could be recovered by reestablishing complementarity in the respective partner strand. Therefore base conversion by RNA editing in GluR-B pre-mRNA is directed by double-stranded RNA formed of exonic and intronic sequences (Higuchi et al. 1993).

The lack of (Q/R) site editing in GluR-A, -C, -D, and GluR7 gene transcripts can be explained by the absence of the requisite intronic ECS elements. For the GluR5 and GluR6 genes the respective *cis*-acting ECS elements are localized approximately 2 kb downstream from the editing site, within the 3' adjacent intron (Herb et al. 1996).

Two further editing sites occur in the M1 segment GluR6 (Köhler et al. 1993). This produces GluR6 subunits with isoleucine (I; codon: ATT) or valine (V; codon: GTT) in one and tyrosine (Y; codon: TAC) or cysteine (C; codon: TGC) in the other M1 position. As a consequence, primary GluR6 transcripts can be edited in three different positions, two in M1 and one in M2. The requisite intronic sequences directing I/V and

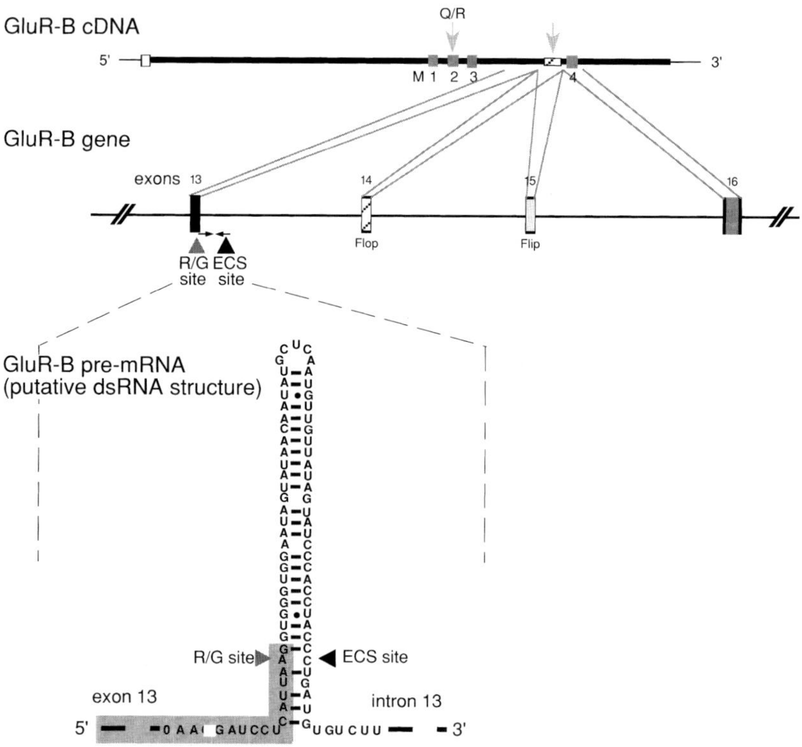

Fig. 3. R/G site editing of the GluR-B pre-mRNA. (Symbols are as in Fig. 2 except that the R/G site is indicated by an *additional arrow*). *Boxes*, exonic sequences of the GluR-B gene: *black*, exon 13; *hatched*, alternatively spliced exons 14 and 15; *gray*, M4 encoding exon 16. *Lines projecting from cDNA to gene*, cDNA portions contributed by exons 13–16. The inverted repeat structure which is necessary for R/G editing is depicted directly after exon 13. *Below*, the putative double-stranded secondary structure which is formed by the inverted repeat sequences of exon 13 and intron 14 of the GluR-B precursor RNA. The nucleotide sequence of the inverted repeat is shown and positions of the to-be-edited R/G-site adenosine and of the ECS are indicated by *gray and black arrowheads*, respectively; *vertical bars between nucleotides*, perfect base pairing; *filled circles*, G-U pairs

Y/C site editing in M1 of GluR6 pre-mRNA need to be delineated. The elecrophysiological analysis of edited and unedited GluR6 subunits recombinantly expressed in HEK293 cells showed that all three positions diversified by RNA editing impact the Ca^{2+} permeability of homomeric GluR6 receptor channels. In the unedited M1 form of GluR6, Ca^{2+} permeability is less dependent on the presence of either Q or R in the M2 region (Köhler et al. 1993).

In addition to residing in receptor segments for membrane insertion, alternative codons, one for arginine (R; AGA) and one for glycine (G; GGA), were also found in the coding region for the S2 segment of the bipartite ligand interacting domain (ref) positioned between M3 and M4 of GluR-B, -C, and GluR-D (Lomeli et al. 1994). This alternative codon is the last codon on exon 13 of the GluR-B gene (Fig. 3) and is spliced alternatively to either the flop (exon 14) or the flip module (exon 15). A survey by reverse-transcription polymerase chain reaction of the relevant coding sequences for the AMPA receptor subunits in adult rat brain demonstrated that flip and flop forms of GluR-B, -C, and -D transcripts occur in both glycine and arginine versions whereas in GluR-A, only the arginine version exists. Thus, except for GluR-A, all AMPA receptor subunit pre-mRNAs undergo nuclear RNA editing to produce R/G variability in the amino acid position preceding the flip and flop splice modules. In the adult rat brain the degree of R/G site editing is variable for the different subunits and ranges from 50% to 95%. During brain development R/G site editing is regulated, starting with a low extent of editing in the embryonic brain followed by dramatic increases for both splice forms until postnatal day 42 (Lomeli et al. 1994). A consistent functional difference between the R and G forms of AMPA receptor channels has been found for recovery rates from desensitization. For example, edited heteromeric GluR-A_{flip}/GluR-C_{flip}(G-form) channels exhibit a faster recovery rate (τ_{rrec}~5 ms) compared to GluR-A_{flip}/GluR-C_{flip}(R-form) channels (τ_{rec}~29 ms). Thus the R/G site constitutes one of the determinants for the time course of recovery from desensitization (Lomeli et al. 1994).

Although mechanistically identical to Q/R site editing, R/G site editing exhibits unique features (Fig. 3). The intronic ECS is located in close vicinity to the exon, directly adjacent to a 24-bp self-complementary intronic sequence which includes the 5' splice donor. The predicted dsRNA structure is approximately 30 bp in length and is imperfect due to mismatches in the three subunit pre-mRNAs. Moreover, the to-be-edited adenosine is mismatched in the exon-intron dsRNA but is base-paired after editing. The reverse is true for Q/R site editing. Also, the 5' splice donor of intron 13 occurs two nucleotides 3' of the edited adenosine while the donor sequence of GluR-B intron 11 is located 20 nucleotides 3' of the to-be-edited adenosine. As editing precedes splicing, interplay between the two posttranscriptional machineries seems likely. This is of particular interest since alternative splicing joins the R/G codon to either the flip or flop exon.

1.5 Candidate Enzymes for GluR Pre-mRNA Editing

Our understanding concerning mechanistic aspects of the site-selective adenosine to guanosine editing in nuclear transcripts is still at an elementary level. Recognizing the involvement of intronic ECSs which form a short intramolecular dsRNA structure with the to-be-edited exonic sequence was a first step in identifying the molecular mechanisms for the adenosine conversions. We had proposed that the to-be-edited adenosines are converted to inosine by adenosine deamination in RNA (Sommer et al. 1991). As a consequence, the preferred base-pairing properties of inosine with cytidine would lead to the incorporation of guanosine in cDNA molecules. The requirement for dsRNA and the fact that only adenosines are targeted in GluR pre-mRNAs have led to the hypothesis that a dsRNA adenosine deaminase catalyzes the nucleotide change (Hough and Bass 1994; Kim et al. 1994; O'Connell et al. 1995). An RNA-dependent adenosine deaminase, termed DRADA or dsRAD, had been known to exist in mammalian tissues for some time, but its function needed elucidation. Previous studies indicated that an in vitro RNA editing system prepared from HeLa cells can be used to determine the nature of the edited adenosine in GluR-B pre-mRNA (reviewed in Bass 1995), and indeed these extracts mediate the site-selective conversion of adenosines to inosines.

Recently, we cloned cDNAs for a dsRNA adenosine deaminase which efficiently edited in vitro the Q/R site in GluR-B pre-mRNA (Melcher et al. 1996). This enzyme, RED1, is predominantly expressed in the brain, but also occurs in peripheral tissues such as testis, lung, and kidney. It is distantly sequence related to the ubiquitously expressed

DRADA which edits the Q/R site poorly. Both RED1 and DRADA deaminate the adenosine in the R/G site of GluR-B pre-mRNA, and hence the two enzymes possess distinct but overlapping substrate specificities. The widespread distribution of the enzymes suggests that other pre-mRNAs, in addition to those of the ionotropic glutamate-receptor family, also undergo site-selective adenosine deamination.

1.6 Early-Onset Epilepsy by Deficiency in GluR-B Q/R Site Editing

To investigate the physiological importance of Q/R site editing in AMPA receptors mice were genetically engineered to incapacitate one of the two GluR-B alleles for RNA editing at the Q/R site (Brusa et al. 1995). This was achieved by removing, in one allele, the ECS from intron 11. Animals heterozygous for the manipulated allele developed seizures and died within 3 weeks of age, demonstrating that GluR-B pre-mRNA editing at the Q/R site is essential for brain function (Fig. 4). Molecular analysis of the GluR-B transcripts in the GluR-$B^{+/\Delta ECS}$ mice revealed that 30% of the GluR-B mRNA occurred in its unedited Q-form. The deviation from the expected 50% unedited GluR-B mRNA (only one of two alleles is incapacitated for Q/R site editing) derives from the fact that the unedited GluR-B pre-mRNA carrying a loxP element in place of the ECS in intron 11 is spliced less efficiently than the edited wild-type pre-mRNA.

The considerable increase in unedited GluR-B pre-mRNA in these mice predicts an increase in the Ca^{2+} permeability of their AMPA receptors. Indeed, measurements in nucleated patches from hippocampal CA1 pyramidal cells, Purkinje cells, neocortical layer 5, and inhibitory dentate gyrus basket cells indicated a shift in the current reversal potential, reflecting a sharply increased Ca^{2+} permeability of AMPA receptors in central neurons of GluR-$B^{+/\Delta ECS}$ mice (Fig. 4). The increased glutamate-activated Ca^{2+} entry into the central excitatory neurons of GluR-$B^{+/\Delta ECS}$ mice did not precipitate detectable changes in brain histology, including the cortical layering of the telencephalon and cerebellum. Myelination was normal and marker molecules such as tyrosine hydroxylase, glutamate decarboxylase, parvalbumin, and calbindin exhibited the same distribution in the brains of GluR-$B^{+/\Delta ECS}$

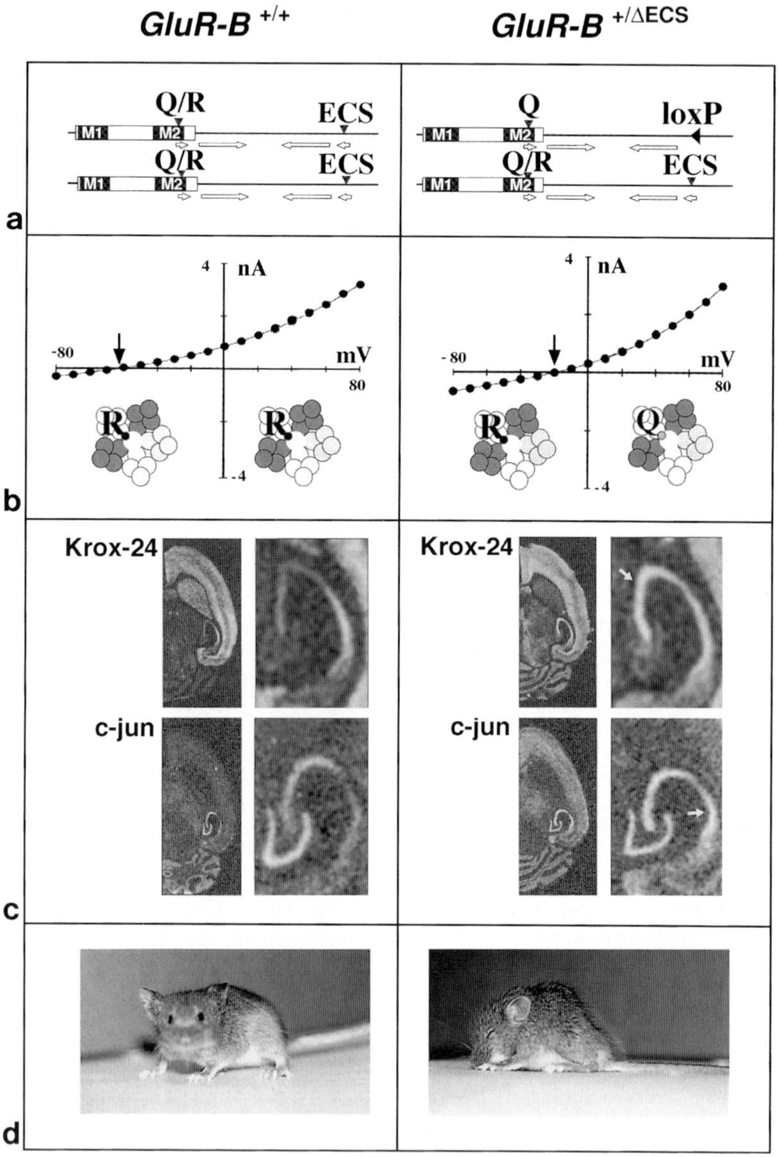

Fig. 4a–d. Legend see p. 13

and GluR-B$^{+/+}$ animals. However, the expression pattern of the Ca^{2+} responsive immediate early genes *Krox-24* and c-*jun* was changed in brains of GluR-B$^{+/\Delta ECS}$ mice relative to GluR-B$^{+/+}$ mice. Starting at P12, all carriers of the GluR-B$^{\Delta ECS}$ allele developed a complex neurological syndrome that included pronounced seizures and death within 3 weeks of age. In the hippocampus, approximately one-half of the pyramidal neurons had died in the lateral aspect of the CA3 region in postmortem GluR-B$^{+/\Delta ECS}$ mice. These findings suggest that onset and severity of the neurological phenotype depends on the ratio of the unedited to edited GluR-B subunits (Q/R ratio) and thus on the increased Ca^{2+} permeability of AMPA receptors in principal neurons.

It is perhaps surprising that a single GluR-B allele rendered imcompetent for Q/R site editing should precipitate early-onset epilepsy and premature lethality. It may be possible to provide a rationale for this dominant effect of the GluR-B$^{\Delta ECS}$ allele, which links RNA editing and epilepsy. Synaptic glutamate-activated Ca^{2+} influx mediated by AMPA receptor channels is high in GABA-ergic interneurons, septal neurons, and dorsal horn spinal neurons, but is low in principal neurons such as pyramidal cells in the hippocampus, neocortical pyramidal cells, and cerebellar Purkinje cells. It is predicted that in interneurons the AMPA receptor-mediated Ca^{2+} influx at the resting membrane potential equals that of synaptically colocalized NMDA receptors and exceeds it at

◀ **Fig. 4a–d.** Early-onset epilepsy and postnatal lethality by an editing deficient GluR-B allele. Schematic representation (**a**) of the GluR-B gene segments encoding the membrane segments M1 and M2 of the GluR-B subunit as present in the wild-type GluR-B$^{+/+}$, and the heterozygous GluR-B$^{+/\Delta ECS}$ mouse (**b**) which was created by gene targeting. Positions of the editing site (Q/R) and the intronic ECS are indicated. *Vertical open arrows*, inverted repeat structures. In the two mouse strains (**b**) gene expression of two Ca^{2+} responsive genes, *Krox-24* and c-*jun*, was monitored at postnatal day 15 by in situ hybridization (**c**) of horizontal brain sections hybridized with ^{35}S-labeled oligonucleotides for *Krox-24* and c-*jun* mRNAs. To permit better evaluation of expression differences hippocampal regions of GluR-B$^{+/+}$ and heterozygote GluR-B$^{+/\Delta ECS}$ mice are enlarged. *Arrows*, CA3 field for *Krox-24* and CA1 field for c-*jun*. **d** The current (nA)/voltage (mV) relations of AMPA receptors in hippocampal CA1 pyramidal cells of GluR-B$^{+/+}$ and GluR-B$^{+/\Delta ECS}$ mice. *Vertical arrows*, the reversal potential in high Ca^{2+} solution which is different in the GluR-B$^{+/+}$ and GluR-B$^{+/\Delta ECS}$ mouse. *Lower right*, a heteromeric AMPAR channel with two GluR-B subunits hyperpolarized states due to the voltage-dependent Mg^{2+} block of NMDA receptors. The Ca^{2+} current conducted by AMPA receptor channels in excitatory synapses of interneurons could shape the postsynaptic currents via Ca^{2+} sensitive K^+ channels or the Ca^{2+}-triggered inactivation of NMDA receptor channels. It might also contribute to long-term changes of synaptic efficacy (Jonas and Burnashev 1995).

In principal neurons, AMPA receptors do not seem to contribute to the generation of a glutamate-elicited Ca^{2+} signal, and hence excitatory synapses in these neurons may be entirely tuned to the processing of NMDA receptor-mediated Ca^{2+} influx which occurs at distinct membrane potentials and with slower time courses. Thus, inefficient editing of the Q/R site in GluR-B pre-mRNA elicits an inappropriate Ca^{2+} signal at excitatory synapses of principal neurons and may lead to an interference in mechanisms optimized for the control of synaptic strength. This in turn may destabilize at the network level, perhaps explaining the epileptic phenotype seen in mice expressing sufficient levels of the unedited GluR-B subunit.

In this context it is important to note that the perturbation of excitatory signaling should correlate with the level of Ca^{2+} influx through synaptically activated AMPA receptor channels rather than with Ca^{2+} permeability as determined from the shift in reversal potential under biionic conditions. This is because AMPA receptors formed without GluR-B and, therefore, possessing maximal Ca^{2+} permeability can mediate only little ion flow across a range of membrane potentials owing to a pronounced current rectification by a voltage-dependent block by intracellular polyamines (Sommer et al. 1991). Thus, a null mutant of GluR-B is predicted to express AMPA receptor channels with very high $P_{Ca}/P_{monovalent}$ permeability ratios, but these channels may not mediate any glutamate-elicited Ca^{2+} flux at depolarized membrane potentials. In fact, synaptic transmission in GluR-B null mice should be functionally impaired due to the strong polyamine block, predicting a pronounced reduction in the excitability of central neurons and hence a phenotype unlikely to be epilepsy prone.

In summary, the Q/R ratio of GluR-B, which results from the extent of Q/R site editing of GluR-B pre-mRNA controls the synaptic influx of Ca^{2+} ions through AMPA receptors. An impairment in the extent of RNA editing at the Q/R site of GluR-B may lead to epilepsy as a direct or indirect consequence of an increased Ca^{2+} influx through AMPA receptors, mainly in principal neurons. Complete deficiency of editing at this site, even though it would increase the Ca^{2+} permeability of AMPA receptors to maximal levels, should subject the conductance of these channels to block by polyamines and, thus, interfere with excitatory neurotransmission.

Acknowledgements. Supported in part by SFB 317 grant B/9, HFSP grant RG-3/95 B, and the German Chemical Industry.

References

Bass BL (1995) An I for editing. Curr Biol 5:598–600

Bernard A, Khrestchatisky M (1994) Assessing the extent of RNA editing in the TMII regions of GluR5 and GluR6 kainate receptors during rat brain development. J Neurochem 62:2057–2060

Bliss TV, Collingridge GL (1993) A synaptic model of memory: long-term potentiation in the hippocampus. Nature 361:31–39

Brusa R, Zimmermann F, Koh D-S, Feldmeyer D, Gass P, Seeburg PH, Sprengel R (1995) Early-onset epilepsy and postnatal lethality associated with an editing-deficient GluR-B allele in mice. Science 270:1677–1680

Burnashev N, Monyer H, Seeburg PH, Sakmann B (1992a) Divalent ion permeability of AMPA receptor channels is dominated by the edited form of a single subunit. Neuron 8:189–198

Burnashev N, Schoepfer R, Monyer H, Ruppersberg JP, Gunther W, Seeburg PH, Sakmann B (1992b) Control by asparagine residues of calcium permeability and magnesium blockade in the NMDA receptor. Science 257:1415–1419

Choi DW (1988) Glutamate neurotoxicity and diseases of the nervous system. Neuron 1:623–634

Collingridge GL, Lester RA (1989) Excitatory amino acid receptors in the vertebrate central nervous system. Pharmacol Rev 41:143–210

Geiger JRP, Melcher T, Koh DS, Sakmann B, Seeburg PH, Jonas P, Monyer H (1995) Relative abundance of subunit mRNAs determines gating and Ca^{2+} permeability of AMPA receptors in principal neurons and interneurons in rat CNS. Neuron 5:193–204

Herb A, Melcher T, Maas S, Sprengel R, Higuchi M, Seeburg PH (1996) Q/R site editing in kainate receptor GluR5 and GluR6 pre-mRNAs requires distant intronic sequences. Proc Natl Acad Sci USA 93:1875–1880

Higuchi M, Single FN, Köhler M, Sommer B, Sprengel R, Seeburg PH (1993) RNA editing of AMPA receptor subunit GluR-B: a base-paired intron-exon structure determines position and efficiency. Cell 75:1361–1370

Hollmann M, Heinemann S (1994) Cloned glutamate receptors. Ann Rev Neurosci 17:31–108
Hollmann M, Hartley M, Heinemann S (1991) Ca^{2+} permeability of KA-AMPA-gated glutamate receptor channels depends on subunit composition. Science 252:851–853
Hollmann M, Maron C, Heinemann S (1994) N-glycosylation site tagging suggests a three transmembrane domain topology for the glutamate receptor GluR1. Neuron 13:1331–1343
Hough RF, Bass BL (1994) Purification of the Xenopus laevis double-stranded RNA adenosine deaminase. J Biol Chem 269:9933–9939
Jonas P, Burnashev N (1995) Molecular mechanisms controlling calcium entry through AMPA-type glutamate receptor channels. Neuron 15:987–990
Kim U, Wang Y, Sanford T, Zeng Y, Nishikura K (1994) Molecular cloning of cDNA for double-stranded RNA adenosine deaminase, a candidate enzyme for nuclear RNA editing. Proc Natl Acad Sci USA 91:11457–11461
Köhler M, Burnashev N, Sakmann B, Seeburg PH (1993) Determinants of Ca^{2+} permeability in both TM1 and TM2 of high affinity kainate receptor channels: diversity by RNA editing. Neuron 10:491–500
Kuner T, Wollmuth LP, Karlin A, Seeburg PH, Sakmann B (1996) Structure of the NMDA receptor channel M2 segment inferred from the accessibility of substituted cysteines. Neuron 17:343–352
Lomeli H, MosbacherJ, Melcher T, Hoger T, Geiger JRP, Kuner T, Monyer H, Higuchi M, Bach A, Seeburg PH (1994) Control of kinetic properties of AMPA receptor channels by nuclear RNA editing. Science 266:1709–1713
Mayer ML, Westbrook GL (1987) The physiology of excitatory amino acids in the vertebrate central nervous system. Prog Neurobiol 28:197–276
Melcher T, Maas S, Herb A, Sprengel R, Seeburg PH, Higuchi M (1996) A mammalian RNA editing enzyme. Nature 379:460–464
Monyer H, Sprengel R, Schoepfer R, Herb A, Higuchi M, Lomeli H, Burnashev N, Sakmann B, Seeburg PH (1992) Heteromeric NMDA receptors: molecular and functional distinction of subtypes. Science 256:1217–1221
Noebels JL (1996) Targeting epilepsy genes. Neuron 16:241–244
O'Connell MA, Krause S, Higuchi M, Hsuan J, Totty NF, Jenny A, Keller W (1995) Cloning of cDNAs encoding mammalian double-stranded RNA-specific adenosine deaminase. Mol Cell Biol 15:1389–1397
Scott J (1995) A place in the world for RNA editing. Cell 81:833–836
Seeburg PH (1996) The role of RNA editing in controlling glutamate receptor channel properties. J Neurochem 66:1–5
Simpson L, Emeson RB (1996) RNA editing. Annu Rev Neurosci 19:27–52
Sommer B, Köhler M, Sprengel R, Seeburg PH (1991) RNA editing in brain controls a determinant of ion flow in glutamate-gated channels. Cell 67:11–19

Sprengel R, Seeburg PH (1994) In: Peroutka SJ (ed) Ionotropic glutamate receptors. Handbook of receptors and channels, vol 2. CRC, Boca Raton, pp 213–264

Stern P, Edwards FA, Sakmann B (1992) Fast and slow components of unitary EPSCs on stellate cells elicited by focal stimuation in slices of rat visual cortex. J Physiol 449:247–278

Watkins JC, Krogsgaard-Larsen P, Honore T (1990) Structure-activity relationships in the development of excitatory amino acid receptor agonists and competitive antagonists. Trends Pharmacol Sci 11:25–33

Wo ZG, Oswald E (1994) Transmembrane topology of two kainate receptor subunits revealed by N-glycosylation. Proc Natl Acad Sci USA 91:7154–7158

2 Regulation of Glutamate Receptor Gene Expression in Global Ischemia

D. E. Pellegrini-Giampietro

2.1 Introduction

Glutamate receptors (GluRs) mediate normal excitatory neurotransmission and play a critical role in a number of other important physiological functions such as synaptogenesis, formation of neuronal circuitry, and synaptic plasticity, including long-term potentiation and long-term depression. However, excessive activation of GluRs contributes to neurodegeneration following a wide range of neurological insults including ischemia, trauma, hypoglycemia, and epileptic seizures. Chronic neurodegenerative disorders such as Alzheimer's disease, Huntington's chorea, AIDS encephalopathy, and amyotrophic lateral sclerosis may also involve glutamate-induced neuronal cell death (Choi 1992; Lipton and Rosenberg 1994). Glutamate promotes neurodegeneration by increasing the concentrations of intracellular free Ca^{2+} in neurons, thereby leading to generation of free radicals and activation of proteases, phospholipases and endonucleases (Siesjö and Bengtsson 1989; Coyle and Puttfarcken 1993), and/or transcriptional activation of specific "cell death" programs (Schreiber and Baudry 1995). Diverse GluRs are known to produce a rise in cytosolic Ca^{2+} by a number of mechanisms including: activation of Ca^{2+}-permeable *N*-methyl-D-aspartate (NMDA) receptors, opening of voltage-dependent Ca^{2+} channels following membrane depolarization induced by α-amino-3-hydroxy-5-methyl-4-isoxazolepropionic acid (AMPA) receptor activation, and activation of metabotropic GluRs linked to phosphoinositide hydrolysis, which releases Ca^{2+} from intracellular stores.

Until recently, AMPA receptors were thought to be Ca^{2+}-impermeable. It is now well established that AMPA receptors lacking the GluR2 subunit are permeable to a number of divalent cations including Ca^{2+}. We have recently proposed a hypothesis (the GluR2 hypothesis) which predicts that specific neurological insults lead to a decrease in GluR2 expression and formation of Ca^{2+}-permeable AMPA receptors, and thereby, enhanced toxicity to endogenous glutamate (Pellegrini-Giampietro et al. 1997). This chapter will focus on recent experimental evidence that Ca^{2+}-permeable AMPA receptors contribute to delayed and cell-specific neurodegeneration following transient global ischemia.

2.2 Molecular Biology of AMPA Receptors

AMPA-type GluRs mediate fast excitatory synaptic transmission in the vertebrate central nervous system. AMPA receptors are ligand-gated channels which are thought to be pentamers assembled from GluR1, 2, 3, and 4 (or GluR-A, -B, -C, and -D) subunits (Nakanishi 1992; Seeburg 1993; Hollmann and Heinemann 1994) around a central aqueous pore. The predicted secondary structure of GluR subunits is depicted in Fig. 1A and includes the following features: (a) A large extracellular

N-terminus domain; (b) three transmembrane-spanning domains (TM1, TM3, and TM4); (c) a fourth hydrophobic segment, (M2), that forms a channel-lining reentrant hairpin loop similar to the pore-forming region of K^+ channels (Wo and Oswald 1995); (d) a binding domain for agonists formed from the S1 and S2 extracellular regions (Stern-Bach et al. 1994); and (e) an intracellular C-terminus domain. GluR subunits appear to be expressed by virtually all neurons in the brain, although each cell may differ in the number and type of subunits expressed. Individual neurons predominantly form heteromeric AMPA receptors made up of at least two different subunits, but they may also form homomers (Martin et al. 1993; Wenthold et al. 1996).

2.2.1 Ca^{2+} Permeability of AMPA Receptors Is Controlled by the GluR2 Subunit

The first demonstrations that GluR2 determines Ca^{2+} permeability of AMPA receptors came from electrophysiological studies of recombinant AMPA channels expressed in *Xenopus laevis* oocytes (Hollmann et al. 1991) and mammalian cells (Verdoorn et al. 1991). AMPA receptors assembled from GluR1, GluR3, and/or GluR4 subunits are permeable to Ca^{2+} and have doubly rectifying current/voltage relations. GluR2, expressed with other GluR subunits, forms channels that are Ca^{2+}-impermeable and electrically linear or outwardly rectifying (see Fig. 1). The dominance of the GluR2 subunit in determining permeability to Ca^{2+} and other divalent ions is attributed to the presence of a positively charged arginine (Fig. 1A; *R*) in place of a glutamine (Fig. 1A; *Q*) residue within the M2 domain (Hume et al. 1991; Burnashev et al. 1992). Rectification of receptors lacking GluR2 arises from fast voltage-dependent channel block by intracellular polyamines (Kamboj et al. 1995; Koh et al. 1995). Some positively charged polyamine spider toxins like argiotoxin (Herlitze et al. 1993) and Joro spider toxin (Blaschke et al. 1993) block Ca^{2+}-permeable AMPA receptors selectively (presumably because they are repelled by the positively charged arginine residue in the GluR2 subunit), and therefore serve as pharmacological probes to detect the presence or absence of GluR2 in recombinant, as well as in native receptor assemblies.

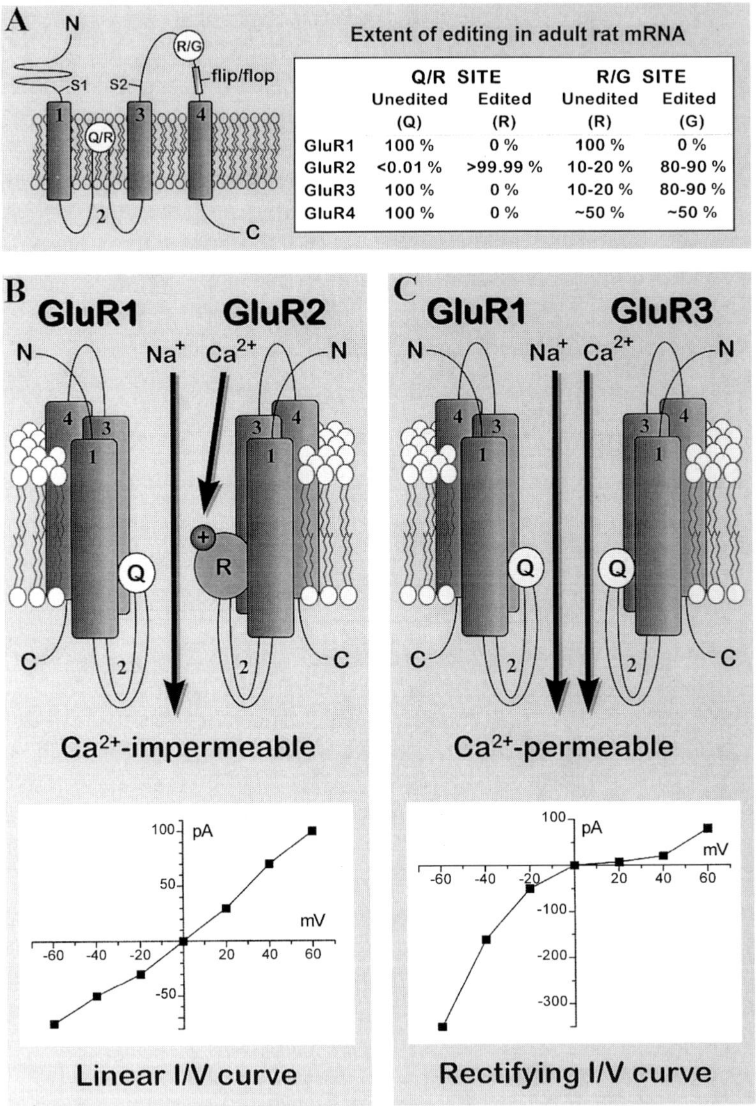

Fig. 1A–C. Legend see p. 23

The functionally critical arginine within M2 is not encoded by the GluR2 gene, but rather arises within the pre-mRNA by editing of a codon for the neutral glutamine residue (Sommer et al. 1991; Higuchi et al. 1993). RNA editing at the Q/R site (see Fig. 1A) is specific to GluR2 and is extremely efficient. In neonatal and adult rat brain, virtually 100% of GluR2 mRNA undergoes editing; in embryonic brain, only a small percentage of GluR2 subunits (about 1%) are unedited (Burnashev et al. 1992) (Fig. 1). More recently, editing at another site was identified in the extracellular TM3-TM4 loop of GluR subunits, which determines a switch from the encoded arginine (Fig. 1A; *R*) to glycine (Fig. 1A; *G*) (Lomeli et al. 1994). Editing at the R/G site is specific for GluR2, -3, and -4, and is about 80%–90% complete in adult rat brain (Fig. 1A). Immediately adjacent to the R/G site, one of two cassettes named *flip* and *flop* (Sommer et al. 1990), each containing 38 amino acids, is

◀ **Fig. 1A–C.** Proposed secondary structure of GluR subunits depicting critical sites conferring functional diversity on α-amino-3-hydroxy-5-methyl-4-isoxazolepropionic acid (AMPA) receptors. **A** A GluR subunit in its linear sequence and inserted into the cell membrane. Features of the predicted GluR-protein structure include: A large extracellular N-terminus domain (*N*); three trans-membrane-spanning domains: TM1 (*1*), TM3 (*3*), and TM4 (*4*); a fourth hydrophobic segment, M2 (*2*), that is thought to make a hairpin turn within the membrane and line the ion channel (Wo and Oswald 1995); two extracellular segments (*S1* and *S2*) that are predicted to form the binding domains for agonists (Stern-Bach et al. 1994); and an intracellular C-terminus domain (*C*). The position of the alternatively spliced flip/flop module and those of edited residues are shown. The alternative amino acids of editing sites are indicated as follows : *R* arginine; *Q*, glutamate; *G*, glycine. The *table* lists the average percentages of unedited and edited GluR2 subunits as evaluated by polymerase chain reaction analysis of adult rat RNA. Data for the Q/R site are from Burnashev et al. (1992) and those for the R/G site from Lomeli et al. (1994). **B** The GluR2 subunit, expressed by *GluR1* and/or *GluR3*, forms recombinant (Hollmann et al. 1991; Verdoorn et al. 1991) or native (Bochet et al. 1994; Jonas et al. 1994; Geiger et al. 1995) AMPA channels that are Ca^{2+}-impermeable and have electrically linear current/voltage relations. GluR2 limits Ca^{2+} gating due to the presence of an edited, positively charged arginine (*R*) in place of a glutamine (*Q*) residue at the Q/R site (Hume et al. 1991; Burnashev et al. 1992). **C** AMPA channels assembled from GluR1 and/or GluR3 subunits are permeable to Ca^{2+} and doubly rectifying. Rectification of receptors lacking GluR2 arises from fast voltage-dependent channel block and permeation at high positive voltages by intracellular polyamines (Kamboj et al. 1995; Koh et al. 1995) introduced in GluR subunits by alternative RNA splicing (Fig. 1A). RNA editing at the R/G site and splicing at the flip/flop site are developmentally regulated and are cooperative in controlling desensitization and recovery rates of AMPA receptor responses (Seeburg 1996). The allosteric modulator cyclothiazide strongly attenuates desensitization in

---

Fig. 2A,B. The GluR2 hypothesis: CA1 pyramidal neurons degenerate but interneurons survive postischemia. **A** During fast excitatory neurotransmission (i.e., brief bouts of normal synaptic activity) glutamate is released and activates α-amino-3-hydroxy-5-methyl-4-isoxazolepropionic acid receptors (*AMPAR*) but not *N*-methyl-D-aspartate receptors (*NMDAR*), which are blocked by extracellular Mg^{2+}. Hippocampal pyramidal cells carry heteromeric AMPA receptors containing the GluR1 (*1*) and GluR2 (*2*) subunits and are therefore permeable only to monovalent cations (Bochet et al. 1994; Geiger et al. 1995). AMPA-mediated Na^{+} influx depolarizes the membrane and allows Ca^{2+} entry via voltage-activated Ca^{2+} channels (*VOCC*); cytosolic free Ca^{2+} is rapidly stored in intracellular compartments or extruded by pumps in the plasma membrane (ATP-dependent Ca^{2+} pumps or the Na^{+}–Ca^{2+} exchanger). γ-Aminobutyric acid (GABA)ergic interneurons carry rapidly desensitizing, Ca^{2+}-permeable, heteromeric AMPA receptors assembled with GluR1 (*1*) and GluR4 (*4*), but lacking GluR2 (Bochet et al. 1994; Geiger et al. 1995; Racca et al. 1996). To deal with relatively abundant intracellular Ca^{2+} levels, interneurons specifically express Ca^{2+}-binding proteins (*CBP*) like parvalbumin, calbindin, and calretinin (Ribak et al. 1990). **B** During an episode of severe global ischemia the release of glutamate increases (*thick line*), but upon reperfusion it returns again to normal levels (Szatkowski and Attwell 1994). The transient, but abundant, outflow of glutamate during ischemia may trigger transcriptional changes in CA1 pyramidal cells resulting in a preferential reduction of GluR2 (*2*) expression which, however, does not appear to be mediated by activation of AMPA or NMDA receptors (Pellegrini-Giampietro et al. 1994). As a result, GluR2-lacking AMPA channels are formed postischemia, either as GluR1 (*1*) homomers or GluR1/3/4 heteromers, gating a rapid and massive influx of Ca^{2+} in response to physiological amounts of glutamate. The ensuing abnormal increase in intracellular free Ca^{2+} exceeds the buffering capacity of pyramidal cells and leads to activation of enzymatic pathways which result in cell death. Although sustained synaptic activity is expected to release the Mg^{2+} block, NMDA receptors are thought to be inactive during severe ischemia because of the high extracellular H^{+} concentrations (Silver and Erecinska 1992). Interneurons do not modify their response to glutamate during or following ischemia and do not degenerate (Johansen et al. 1983), not only because they are used to handling a continuous influx of Ca^{2+}, but also because anoxia preferentially blocks the excitatory input to GABAergic interneurons (Khazipov et al. 1993; Congar et al. 1995). This functional disconnection also leads to further excitation of pyramidal cells ▶

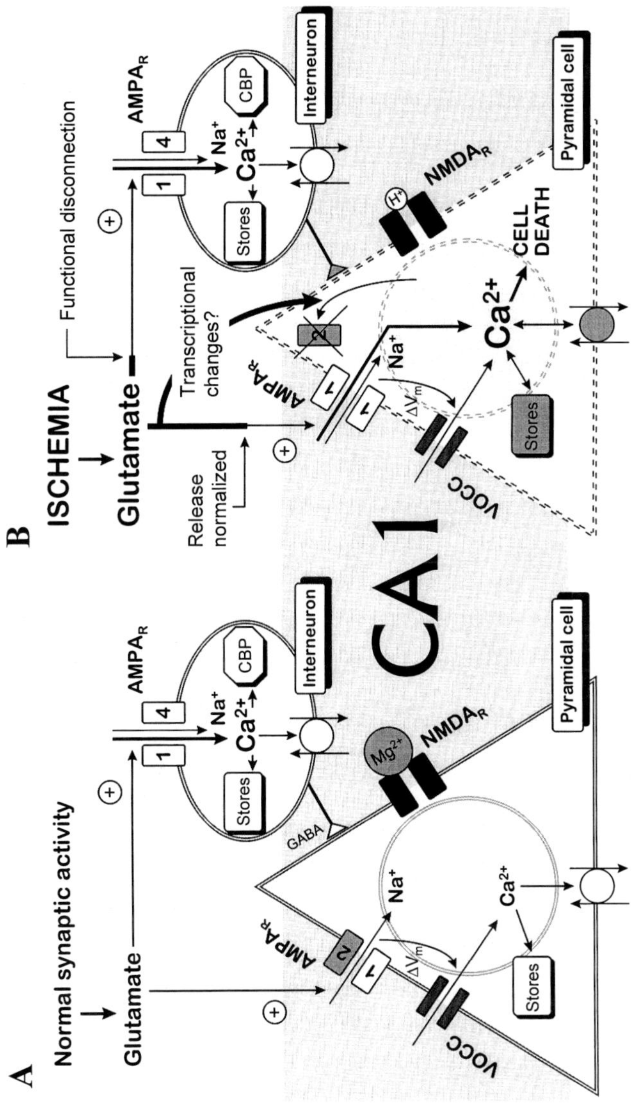

Fig. 2A,B. Legend see p. 24

flip, but not in flop, splice variants of recombinant AMPA receptors (Partin et al. 1994). Consistent with their extracellular position, neither the R/G site nor the flip/flop cassettes affect Ca^{2+} permeability of AMPA receptor channels.

2.2.2 Ca^{2+}-Permeable AMPA Receptors in Native Neurons: A Mechanism for Neurodegeneration?

The Ca^{2+} permeability of native AMPA receptors varies widely (for reviews see Jonas and Burnashev 1995; Burnashev 1996). Since editing of GluR2 mRNA at the Q/R site is virtually complete under physiological conditions, Ca^{2+}-permeable AMPA receptors arise in neurons only as a consequence of reduced expression of GluR2 mRNA. Studies involving patch-clamp recording and reverse transcriptase polymerase chain reaction (RT-PCR) demonstrate that AMPA receptor permeability to Ca^{2+} varies inversely with abundance of GluR2 mRNA in a wide range of cell types. Excitatory principal neurons such as hippocampal (Bochet et al. 1994; Geiger et al. 1995) and neocortical (Jonas et al. 1994) pyramidal cells and dentate gyrus granule cells (Geiger et al. 1995) exhibit low Ca^{2+} permeability and more abundant GluR2 mRNA. Hippocampal (Bochet et al. 1994; Racca et al. 1996) and neocortical (Jonas et al. 1994) γ-aminobutyric acid (GABA)ergic interneurons and dentate gyrus basket cells (Geiger et al. 1995) display higher Ca^{2+} permeability and less abundant GluR2 mRNA. Bergmann glia cells of the cerebellum exhibit high Ca^{2+} permeability and no detectable GluR2 (Geiger et al. 1995).

The GluR2 hypothesis predicts that Ca^{2+}-permeable AMPA receptors may lead to neurotoxic events. However, a number of reports would appear to contradict this view. For example, hippocampal GABAergic interneurons lacking GluR2 are viable and relatively resistant to postischemic delayed neurodegeneration (Johansen et al. 1983). Moreover, nitric oxide synthase-positive neurons of cortex, striatum, and hippocampus, sharing the common feature of low GluR2 expression (Catania et al. 1995), are relatively resistant in neurodegenerative diseases. Finally, transgenic mice with targeted disruption of the GluR2 gene survive and their principal neurons (that express abundant levels of GluR2 in wild type animals) are functional (Jia et al. 1996). Possible explanations for the survival of neurons that normally lack GluR2 may include compensatory mechanisms for Ca^{2+} buffering and extrusion [as, for example, expression of Ca^{2+}-binding proteins (Ribak et al. 1990)], or reduced AMPA currents due to expression of receptors with faster and more profound desensitization (Geiger et al. 1995; Lambolez et al. 1996). Hence, the GluR2 hypothesis will apply primarily to neurons that normally express Ca^{2+}-impermeable AMPA receptors and that are not required to cope with Ca^{2+} influx via AMPA receptors. In these cells, acute increases in permeability of AMPA receptors to Ca^{2+} could account for cell death (Fig. 2). A high rate of Ca^{2+} influx through AMPA receptors in cultured rat cerebellar Purkinje cells (Brorson et al. 1994), in a subpopulation of neocortical neurons (Turetsky et al. 1994; Weiss et al. 1994; Lu et al. 1996), and in an immortalized rat oligodendroglial cell line (Yoshioka et al. 1995) leads to enhanced vulnerability to agonist-induced cytotoxicity. Heterozygous transgenic mice engineered for a Q/R editing deficient GluR2 allele (Brusa et al. 1995) express AMPA receptors with increased Ca^{2+} permeability, particularly in hippocampal and neocortical principal neurons. The mice develop recurrent seizures and die within the first 3 weeks of life, with cell loss in the hippocampus. In these animals, unedited GluR2 may contribute to the formation of a greater number of Ca^{2+}-permeable AMPA receptors than in the GluR2 knockout mice. There are other examples where the phenotypes of different knockout mutants of similar design are unexpectedly different, ranging from complete viability to lethality (Olson et al. 1996).

2.3 Ca^{2+}-Permeable AMPA Receptors in Global Ischemia

Global or forebrain ischemia is characterized by the transient reduction or elimination of blood flow to both cerebral hemispheres. In rat (Pulsinelli et al. 1982) and gerbil (Kirino 1982) models of severe global ischemia and in patients successfully resuscitated from cardiorespiratory arrest (Petito et al. 1987), all forebrain areas are equally affected by oxygen and glucose deprivation, but only selected neuronal populations degenerate and die (for review, see Schmidt-Kastner and Freund 1991). Pyramidal cells in the CA1 subfield of the hippocampus are particularly vulnerable. However, histological evidence of neurodegeneration that exhibits characteristics of apoptosis (Héron et al. 1993; MacManus et al. 1995; Nitatori et al. 1995, but see Deshpande et al. 1992), is not observed until 48–72 h after circulation has been restored. This delayed neurodegeneration, which may have clinical relevance, is thought to be triggered by a transient rise in glutamate release during the ischemic episode (Szatkowski and Attwell 1994), followed by late and excessive Ca^{2+} influx through GluR channels (Siesjö and Bengtsson 1989; Choi 1995). Although NMDA receptors are highly Ca^{2+}-permeable, there is now general consensus that antagonists of AMPA receptors, like the quinoxalinedione NBQX, appear to be much more effective than NMDA antagonists in preventing CA1 cell death following severe global ischemia, even when given as late as 24 h after ischemia (Buchan et al. 1991; Nellgard and Wieloch 1992; Sheardown et al. 1993). These observations indicate that AMPA receptor activation is necessary, and possibly sufficient, for delayed postischemic degeneration and that cells are not irreversibly damaged until well after the ischemic episode.

Fig. 3A–C. Expression of the GluR2 subunit is selectively depressed in CA1 24 h postischemia in rats. In each panel, values are plotted as percent of control mRNA levels reported at time 0 ± standard error of mean. **A** Hybridization levels in CA1 decrease for all probes between 6–12 h. GluR1 and GluR3 did not significantly change between 12–24 h, whereas GluR2 continued to decrease. At 24 h GluR2 is significantly different vs. GluR1 and GluR3. **B** Hybridization levels in CA3. Modest decreases are observed for all probes. GluR2 is significantly different at 24 h vs. earlier time points but not vs. GluR1 or GluR3. **C** Hybridization levels in dentate gyrus. None of the probes are significantly different from controls at any time point. Statistical analysis was performed on logit-transformed normalized percent optical densities by analysis of variance and Tukey's *w*-test for multiple comparisons. *a*=$p<0.05$ vs. 1 h; *b*=$p<0.05$ vs. 1 h or 6 h; *c*=$p<0.05$ vs. 18 h and vs. GluR3 at 24 h and $p<0.01$ vs. 12 h and vs. GluR1 at 24 h; *d*=$p<0.05$ vs. 1 h and $p<0.01$ vs. 6 h. $n=3$ (1 h and 6 h); $n=4$ (controls, 12 h, 18 h, and 24 h). *DG*, dentate gyrus. [Reprinted with permission from Pellegrini-Giampietro et al. (1992b)] ▶

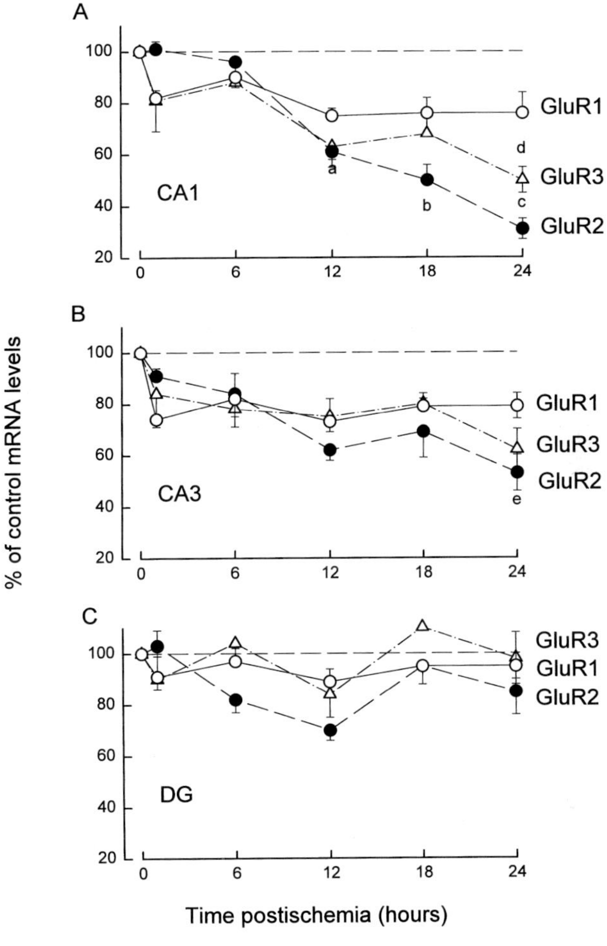

Fig. 3A–C. Legend see p. 28

2.3.1 GluR2 mRNA Is Reduced in Vulnerable Cells Following Global Ischemia

Transient forebrain ischemia in the rat induces an intriguing modification in expression of AMPA receptor subunit mRNAs, as detected by in situ hybridization (Pellegrini-Giampietro et al. 1992b). GluR2 mRNA expression is markedly reduced in CA1 (the brain region most vulnerable to ischemia-induced damage), but not in CA3 and dentate gyrus. Analysis of emulsion-dipped sections indicates that the reduction is specific to CA1 pyramidal cells, and reveals that it clearly precedes any histological sign of cell damage. The decline in GluR2 mRNA in CA1 is first detectable at about 12 h after ischemia, and is about 30% of control at 24 h. GluR3 expression is also reduced, but less so (to about 50% of control), whereas GluR1 and NR1 mRNA expression are not significantly changed (Fig. 3). The (GluR1+GluR3)/GluR2 ratio, a possible predictor of formation of Ca^{2+}-permeable AMPA receptors (Pellegrini-Giampietro et al. 1992a), is markedly increased in CA1 at 24 h, but not in CA3 or dentate gyrus. Because the GluR2 subunit limits the Ca^{2+} permeability of AMPA channels, these findings predict increased formation of Ca^{2+}-permeable AMPA receptors in CA1 neurons after transient global ischemia at times preceding neurodegeneration. Moreover, timing of the switch is consistent with a causal role for AMPA receptor-mediated Ca^{2+} influx in postischemic damage.

Although Frank et al. (1995), using similar conditions in the rat, could not reproduce these findings, subsequent studies from our laboratory in rats (Pellegrini-Giampietro et al. 1994) and gerbils (Gorter et al. 1997) have repeated the observation of a selective decline in expression of GluR2 (as compared to GluR1, GluR3, and the NMDA receptor subunit NR1) in CA1 following global ischemia and preceding neurodegeneration. In addition, two other groups (Pollard et al. 1993; Heurteaux et al. 1995) have reported that the GluR2 subunit is markedly depressed 24 h following transient forebrain ischemia in the same vulnerable region and not in resistant areas, both flip and flop variants being reduced. Unfortunately, these studies provide no information on other GluR subunits (namely GluR1 and GluR3) that could be forming Ca^{2+}-permeable AMPA channels in CA1 prior to their degeneration. Interestingly, the Q/R editing process inserting a crucial arginine site in GluR2 mRNA, which is another mechanism capable of modulating Ca^{2+} fluxes via AMPA receptors, does not appear to be altered 24 h following transient global ischemia (Kamphuis et al. 1995; Paschen et al. 1996; Rump et al. 1996).

2.3.2 Ca^{2+}-Permeable AMPA Receptors Are Formed in Vulnerable Cells Following Global Ischemia

Although an inverse correlation has been established between Ca^{2+} permeability of AMPA receptors and abundance of GluR2 mRNA in normal neurons (Jonas and Burnashev 1995; Burnashev 1996), this relation may not hold under pathological conditions. Thus, functional studies are necessary to establish whether CA1 AMPA receptors in pyramidal cells become permeable to Ca^{2+} in postischemic brain. Early studies involving measurement of $^{[45]}Ca^{2+}$ uptake or Ca^{2+} content showed that Ca^{2+} levels in the CA1 of gerbils (Dux et al. 1987) and rats (Dienel 1984; Deshpande et al. 1987) are maximal at 24–72 h postischemia. A subsequent study by Silver and Erecinska (1992) reported a late rise in intracellular Ca^{2+} concentrations in CA1 neurons following global ischemia but preceding degeneration; this elevation does not appear to involve entry through L-type voltage-dependent channels or NMDA receptors.

More recently, electrophysiological and optical imaging studies using hippocampal slices from gerbils at various times following global ischemia have focused on AMPA receptor-mediated Ca^{2+} influx. Tsubokawa and colleagues (1994) showed that electrically evoked excitatory postsynaptic currents (EPSCs) of gerbil CA1 neurons (1.5–3 days following ischemia) exhibited a prolonged AMPA-mediated component, defined by its resistance to the NMDA blocker AP5. The prolongation was reversed by lowering external Ca^{2+} or injecting the Ca^{2+} buffer 1,2-bis(2-aminophenoxy)ethane-N,N,N',N'-tetraacetic acid (BAPTA) into the CA1 neuron. Part of the AMPA-mediated component was blocked by Joro spider toxin (Tsubokawa et al. 1995) (which is selective for GluR2-less, Ca^{2+}-permeable AMPA receptors). These findings indicate that postischemic CA1 pyramidal cells have abnormal EPSCs, a component of which is mediated by Ca^{2+}-permeable AMPA receptors.

To examine whether GluR2 downregulation leads to AMPA receptor-mediated Ca^{2+} influx in CA1 neurons postischemia, a combined electrophysiological and optical imaging approach was applied to hippocampal slices from postischemic gerbils (Gorter et al. 1997). Parallel in situ hybridization indicated that GluR2 mRNA expression was markedly and specifically reduced in CA1 pyramidal cells following global ischemia, but with a slower onset than in rats. Optical ratio imaging with the Ca^{2+} indicator dye fura-2 revealed that AMPA induced a transient but pronounced increase in intracellular free Ca^{2+} in individual CA1 neurons 72 h after the ischemic insult, but not in controls. The AMPA receptor-mediated Ca^{2+} influx was blocked by CNQX; basal intracellular free Ca^{2+} concentrations did not differ in control and postischemic pyramidal cells. These findings indicate that AMPA receptor functional responses are altered following global ischemia and provide evidence for Ca^{2+} influx directly through AMPA receptors in vulnerable CA1 postischemic neurons at times preceding obvious cell loss. Recently, in dissociated rat hippocampal neurons, sublethal oxygen-glucose deprivation induced an increase in GluR4 mRNA expression relative to other AMPA receptor subunits (Ying et al. 1997). Conditioned neurons also exhibited an increase in kainate-induced, Joro toxin-sensitive Ca^{2+} influx, suggesting that the increase in GluR4 was associated with decreased expression of GluR2. In additon, susceptibility to kainate-induced cell death was enhanced.

If Ca^{2+} permeability of AMPA receptors is important in causing neurodegeneration following global ischemia, an important question to be answered is why hippocampal GABAergic interneurons, that are known to express almost exclusively Ca^{2+}-permeable AMPA channels lacking GluR2 (Bochet et al. 1994; Racca et al. 1996), are relatively resistant to ischemic injury (Johansen et al. 1983). Reasons for the different fate of these two cell types may include: (a) Faster time course and increased extent of AMPA receptor desensitization in interneurons (Geiger et al. 1995; Lambolez et al. 1996); (b) more efficient Ca^{2+} sequestration and/or buffering mechanisms in interneurons, possibly due to specific expression of Ca^{2+}-binding proteins (Ribak et al. 1990); (c) functional disconnection of interneurons from excitatory inputs during ischemia (Khazipov et al. 1993; Congar et al. 1995); and (d) massive and acute (rather than continuous) Ca^{2+} influx in pyramidal cells postischemia (see Fig. 2).

2.3.3 Mechanisms for Altered Gene Expression Following Global Ischemia

The molecular mechanisms underlying regulation of GluR2 expression postischemia are as yet unknown. A number of diverse strategies and drug treatments that afford neuroprotection in animal models of global ischemia prevent downregulation of GluR2 expression in CA1. These include preconditioning with a sublethal ischemic episode and administration of hyperpolarizing agents (i.e., activators of ATP-sensitive K^+ channels or adenosine A1 receptor agonists) in rats (Heurteaux et al. 1995), or i.c.v. injection of the endonuclease inhibitor aurintricarboxylic acid in gerbils (E. Aronica and R. S. Zukin, personal communication). In contrast, the AMPA receptor antagonist NBQX given at the time of the ischemic episode or later does not prevent the reduction in GluR2 (Pellegrini-Giampietro et al. 1994), suggesting that NBQX affords neuroprotection by directly blocking modified AMPA channels (i.e., NBQX is likely to remain in the system long enough to act on the newly formed GluR2-less, Ca^{2+}-permeable receptors). The NMDA antagonist MK-801 is not neuroprotective, given early or late, and does not prevent the downregulation of GluR2 (Pellegrini-Giampietro et al. 1994). These data suggest that the GluR2 reduction postischemia is not triggered by activation of ionotropic GluRs neither during the early rise in extracellular glutamate nor later after ischemia [but see Garthwaite and Garthwaite (1989), who propose a role for metabotropic GluRs in triggering delayed neurodegeneration]. The findings with hyperpolarizing agents suggest that increased extracellular K^+ concentration during ischemia (Martin et al. 1994) is an early mechanism that may trigger downregulation of GluR2. Cultured cerebellar granule cells grown in medium containing increasing concentrations of KCl exhibit decreased GluR2 mRNA and protein expression and increased AMPA-evoked $^{[45]}Ca^{2+}$ influx (Condorelli et al. 1993).

After ischemia, expression of transcription factors, including products of immediate early genes, stress proteins, and neurotrophic factors are also altered in CA1 (Nowak et al. 1993; Lindvall et al. 1994). These proteins are plausible candidates for the downregulation of GluR2 expression by reducing mRNA transcription and/or stability. Since some mRNAs and proteins are upregulated, the decrease in GluR2 is not simply due to a loss of transcriptional or translational capability but appears to result from a regulatory, although maladaptive mechanism. Increased Ca^{2+} influx through GluR2-less AMPA receptors in response to endogenous glutamate is then expected to lead to oxidative stress and possibly activation of apoptotic genes resulting in delayed postischemic neurodegeneration.

2.4 Conclusions

This chapter has reviewed evidence that severe global ischemia triggers a molecular switch that shuts off GluR2 AMPA receptor gene expression in cells destined to die. The GluR2 hypothesis predicts that Ca^{2+} entry through GluR2-lacking AMPA receptors in neurons that normally express Ca^{2+}-impermeable channels contributes to or causes delayed cell death in response to endogenous glutamate. In addition to their role in neurological disorders, Ca^{2+}-permeable AMPA receptors are thought to serve a number of physiological functions, including strengthening of synaptic transmission in certain spinal neurons (Gu et al. 1996) and possibly activation of Ca^{2+}-dependent K^+ channels and inactivation of NMDA receptors (see Jonas and Burnashev 1995; Burnashev 1996). Increased Ca^{2+} permeation through AMPA receptors may also be involved in the development of kindled seizures (Prince et al. 1995). The underlying mechanism by which Ca^{2+} permeability is increased comprises, in all physiological and pathological cases reported to date, a reduction in GluR2 gene expression rather than decreased editing at the Q/R site (Burnashev et al. 1992; Kamphuis and Lopes da Silva 1995; Kamphuis et al. 1995; Paschen et al. 1996; Rump et al. 1996). Reduction in GluR2 expression has also been observed in other vulnerable neuronal populations including CA3 pyramidal cells following kainate-induced status epilepticus in rats (Friedman et al. 1994), cerebellar Purkinje and granule cells in mutant spastic rats (Margulies et al. 1993), pyramidal cells of the parahippocampal gyrus in schizophrenia (Eastwood et al. 1995), and spinal motor neurons in amyotrophic lateral sclerosis (Virgo et al. 1996). These examples indicate the more general applicability of the GluR2 hypothesis to other neuropsychiatric diseases and disorders involving glutamate-induced cell death. If the underlying mechanism is proven correct, there is promise in the development of antagonists specific for AMPA receptors, in particular those that are

Ca^{2+}-permeable, for intervention in the neurodegeneration associated with stroke and cardiac arrest.

References

Blaschke M, Keller BU, Rivosecchi R, Hollmann M, Heinemann S, Konnerth A (1993) A single amino acid determines the subunit-specific spider toxin block of α-amino-3-hydroxy-5-methylisoxazole-4-propionate/kainate receptor channels. Proc Natl Acad Sci USA 90:6528–6532

Bochet P, Audinat E, Lambolez B, Crepel F, Rossier J, Iino M, Tsuzuki K, Ozawa S (1994) Subunit composition at the single-cell level explains functional properties of a glutamate-gated channel. Neuron 12:383–388

Brorson JR, Manzolillo PA, Miller RJ (1994) Ca^{2+} entry via AMPA/KA receptors and excitotoxicity in cultured cerebellar Purkinje cells. J Neurosci 14:187–197

Brusa R, Zimmermann F, Koh D-S, Feldmeyer D, Gass P, Seeburg PH, Sprengel R (1995) Early-onset epilepsy and postnatal lethality associated with an editing-deficient GluR-B allele in mice. Science 270:1677–1680

Buchan AM, Li H, Cho S, Pulsinelli WA (1991) Blockade of AMPA receptor prevents CA1 hippocampal injury following severe but transient forebrain ischemia in adult rats. Neurosci Lett 132:255–258

Burnashev N (1996) Calcium permeability of glutamate-gated channels in the central nervous system. Curr Opin Neurobiol 6:311–317

Burnashev N, Monyer H, Seeburg PH, Sakmann B (1992) Divalent ion permeability of AMPA receptor channels is dominated by the edited form of a single subunit. Neuron 8:189–198

Catania MV, Tölle TR, Monyer H (1995) Differential expression of AMPA receptor subunits in NOS-positive neurons of cortex, striatum, and hippocampus. J Neurosci 15:7046–7061

Choi DW (1992) Bench to bedside: the glutamate connection. Science 258:241–243

Choi DW (1995) Calcium: still center-stage in hypoxic-ischemic neuronal death. Trends Neurosci 18:58–60

Condorelli DF, Dell'Albani P, Aronica E, Genazzani AA, Casabona G, Corsaro M, Balazs R, Nicoletti F (1993) Growth conditions differentially regulate the expression of α-amino-3-hydroxy-5-methylisoxazole-4-propionate (AMPA) receptor subunits in cultured neurons. J Neurochem 61:2133–2139

Congar P, Khazipov R, Ben-Ari Y (1995) Direct demonstration of functional disconnection by anoxia of inhibitory interneurons from excitatory inputs in rat hippocampus. J Neurophysiol 73:421–426

Coyle JT, Puttfarcken P (1993) Oxidative stress, glutamate and neurodegenerative disorders. Science 262:689–695
Deshpande JK, Siesjö BK, Wieloch T (1987) Calcium accumulation and neuronal damage in the rat hippocampus following cerebral ischemia. J Cereb Blood Flow Metab 7:89–95
Deshpande JK, Bergstedt K, Lindén T, Kalimo H, Wieloch T (1992) Ultrastructural changes in the hippocampal CA1 region following transient cerebral ischemia: evidence against programmed cell death. Exp Brain Res 88:91–105
Dienel GA (1984) Regional accumulation of calcium in postischemic rat brain. J Neurochem 43:913–925
Dux E, Mies G, Hossmann K-A, Siklos L (1987) Calcium in the mitochondria following brief ischemia of gerbil brain. Neurosci Lett 78:295–300
Eastwood SL, McDonald B, Burnet PWJ, Beckwith JP, Kerwin RW, Harrison PJ (1995) Decreased expression of mRNAs encoding non-NMDA glutamate receptors GluR1 and GluR2 in medial temporal lobe neurons in schizophrenia. Mol Brain Res 29:211–223
Frank L, Diemer NH, Kaiser F, Sheardown M, Rasmussen JS, Kristensen P (1995) Unchanged balance between levels of mRNA encoding AMPA glutamate receptor subtypes following global cerebral ischemia in the rat. Acta Neurol Scand 92:337–343
Friedman LK, Pellegrini-Giampietro DE, Sperber EF, Bennett MVL, Moshé SL, Zukin RS (1994) Kainate-induced status epilepticus alters glutamate and $GABA_A$ receptor gene expression in adult rat hippocampus: an in situ hybridization study. J Neurosci 14:2697–2707
Garthwaite G, Garthwaite J (1989) Quisqualate neurotoxicity: a delayed, CNQX-sensitive process triggered by a CNQX-insensitive mechanism in young rat hippocampal slices. Neurosci Lett 99:113–118
Geiger JRP, Melcher T, Koh D-S, Sakmann B, Seeburg PH, Jonas P, Monyer H (1995) Relative abundance of subunit mRNAs determines gating and Ca^{2+} permeability of AMPA receptors in principal neurons and interneurons in rat CNS. Neuron 15:193–204
Gorter JA, Petrozzino JJ, Aronica E, Rosenbaum D, Opitz T, Bennett MVL, Connor JA, Zukin RS (1997) Global ischemia induces downregulation of GluR2 mRNA and increases AMPA receptor-mediated Ca^{2+} influx in hippocampal CA1 neurons of gerbil. J Neurosci 17:6179–6188
Gu JG, Albuquerque C, Lee CJ, MacDermott AB (1996) Synaptic strengthening through activation of Ca^{2+}-permeable AMPA receptors. Nature 381:793–796
Herlitze S, Raditsch M, Ruppersberg JP, Jahn W, Monyer H, Schoepfer R, Witzemann V (1993) Argiotoxin detects molecular differences in AMPA receptor channels. Neuron 10:1131–1140

Héron A, Pollard H, Dessi F, Moreau J, Lasbennes F, Ben-Ari Y, Charriaut-Marlangue C (1993) Regional variability in DNA fragmentation after global ischemia evidenced by combined histological and gel electrophoresis observations in the rat brain. J Neurochem 61:1973–1976

Heurteaux C, Lauritzen I, Widmann C, Lazdunski M (1995) Essential role of adenosine, adenosine A1 receptors, and ATP-sensitive K^+ channels in cerebral ischemic preconditioning. Proc Natl Acad Sci USA 92:4666–4670

Higuchi M, Single FN, Köhler M, Sommer B, Sprengel R, Seeburg PH (1993) RNA editing of AMPA receptor subunit GluR-B: a base-paired intron-exon structure determines position and efficiency. Cell 75:1361–1370

Hollmann M, Heinemann S (1994) Cloned glutamate receptors. Annu Rev Neurosci 17:31–108

Hollmann M, Hartley M, Heinemann S (1991) Ca^{2+} permeability of KA-AMPA-gated glutamate receptor channels depends on subunit composition. Science 252:851–853

Hume R, Dingledine R, Heinemann S (1991) Identification of a site in glutamate receptor subunits that controls calcium permeability. Science 253:1028–1031

Jia Z, Agopyan N, Miu P, Xiong Z, Henderson J, Gerlai R, Taverna FA, Velumian A, MacDonald J, Carlen P, Abramow-Newerly W, Roder J (1996) Enhanced LTP in mice deficient in the AMPA receptor GluR2. Neuron 17:945–956

Johansen FF, Jorgensen MB, Diemer NH (1983) Resistance of hippocampal CA1 interneurons to 20 min of transient cerebral ischemia in the rat. Acta Neuropathol (Berl) 61:135–140

Jonas P, Burnashev N (1995) Molecular mechanisms controlling calcium enrty through AMPA-type glutamate receptor channels. Neuron 15:987–990

Jonas P, Racca C, Sakmann B, Seeburg PH, Monyer H (1994) Differences in Ca^{2+} permeability of AMPA-type glutamate receptor channels in neocortical neurons caused by differential GluR-B subunit expression. Neuron 12:1281–1289

Kamboj SK, Swanson GT, Cull-Candy SG (1995) Intracellular spermine confers rectification on rat calcium-permeable AMPA and kainate receptors. J Physiol (Lond) 486:297–303

Kamphuis W, Lopes da Silva FH (1995) Editing status of the Q/R site of glutamate receptor-A, -B, -5 and -6 subunit mRNA in the hippocampal kindling model of epilepsy. Mol Brain Res 29:35–42

Kamphuis W, de Leeuw F-E, Lopes da Silva FH (1995) Ischaemia does not alter the editing status at the Q/R site of glutamate receptor-A, -B, -5 and -6 subunit mRNA. Neuroreport 6:1133–1136

Khazipov R, Bregestovski P, Ben-Ari Y (1993) Hippocampal inhibitory interneurons are functionally disconnected from excitatory inputs by anoxia. J Neurophysiol 70:2251–2259

Kirino T (1982) Delayed neuronal death in the gerbil hippocampus following ischemia. Brain Res 239:57–69

Koh D-S, Burnashev N, Jonas P (1995) Block of native Ca^{2+}-permeable AMPA receptors in rat brain by intracellular polyamines generates double rectification. J Physiol (Lond) 486:305–312

Lambolez B, Ropert N, Perrais D, Rossier J, Hestrin S (1996) Correlation between kinetics and RNA splicing of α-amino-3-hydroxy-5-methylisoxazole-4-propionic acid receptors in neocortical neurons. Proc Natl Acad Sci USA 93:1797–1802

Lindvall O, Kokaia Z, Bengzon J, Elmér E, Kokaia M (1994) Neurotrophins and brain insults. Trends Neurosci 17:490–496

Lipton SA, Rosenberg PA (1994) Excitatory amino acids as a final common pathway for neurological disorders. N Engl J Med 330:613–622

Lomeli H, Mosbacher J, Melcher T, Höger T, Geiger JRP, Kuner T, Monyer H, Higuchi M, Bach A, Seeburg PH (1994) Control of kinetic properties of AMPA receptor channels by nuclear RNA editing. Science 266:1709–1713

Lu YM, Yin HZ, Chiang J, Weiss JH (1996) Ca^{2+}-permeable AMPA/kainate and NMDA channels: high rate of Ca^{2+} influx underlies potent induction of injury. J Neurosci 16:5457–5465

MacManus JP, Hill IE, Preston E, Rasquinha I, Walker T, Buchan AM (1995) Differences in DNA fragmentation following transient cerebral or decapitation ischemia in rats. J Cereb Blood Flow Metab 15:728–737

Margulies JE, Cohen RW, Levine MS, Watson JB (1993) Decreased GluR2(B) receptor subunit mRNA expression in cerebellar neurons at risk for degeneration. Dev Neurosci 15:110–120

Martin LJ, Blackstone CD, Levey AI, Huganir RL, Price DL (1993) AMPA glutamate receptor subunits are differentially distributed in rat brain. Neuroscience 53:327–358

Martin RL, Lloyd HGE, Cowan AI (1994) The early events of oxygen and glucose deprivation: setting the scene for neuronal death? Trends Neurosci 17:251–257

Nakanishi S (1992) Molecular diversity of glutamate receptors and implications for brain function. Science 258:597–603

Nellgard B, Wieloch T (1992) Post ischemic blockade of AMPA but not NMDA receptors mitigates neuronal damage in the rat brain following transient severe forebrain ischemia. J Cereb Blood Flow Metab 12:1–11

Nitatori T, Sato N, Waguri S, Karasawa Y, Araki H, Shibanai K, Kominami E, Uchiyama Y (1995) Delayed neuronal death in the CA1 pyramidal cell layer of the gerbil hippocampus following transient ischemia is apoptosis. J Neurosci 15:1001–1011

Nowak TSJ, Osborne OC, Suga S (1993) Stress protein and proto-oncogene expression as indicators of neuronal pathophysiology after ischemia. Prog Brain Res 96:195–208

Olson EM, Arnold H-H, Rigby PWJ, Wold BJ (1996) Know your neighbors: three phenotypes in null mutants of the myogenic bHLH gene MRF4. Cell 85:1–4

Partin KM, Patneau DK, Mayer ML (1994) Cyclothiazide differentially modulates desensitization of AMPA receptor splice variants. Mol Pharmacol 46:129–138

Paschen W, Schmitt J, Uto A (1996) RNA editing of glutamate receptor subunits GluR2, GluR5, and GluR6 in transient cerebral ischemia in the rat. J Cereb Blood Flow Metab 16:548–556

Pellegrini-Giampietro DE, Bennett MVL, Zukin RS (1992a) Are Ca^{2+}-permeable AMPA/kainate receptors more abundant in the immature brain? Neurosci Lett 144:65–69

Pellegrini-Giampietro DE, Zukin RS, Bennett MVL, Cho S, Pulsinelli WA (1992b) Switch in glutamate receptor subunit gene expression in CA1 subfield of hippocampus following global ischemia in rats. Proc Natl Acad Sci USA 89:10499–10503

Pellegrini-Giampietro DE, Pulsinelli WA, Zukin RS (1994) NMDA and non-NMDA receptor gene expression following global ischemia in rats: effect of NMDA and non-NMDA receptor antagonists. J Neurochem 62:1067–1073

Pellegrini-Giampietro DE, Gorter JA, Bennett MVL, Zukin RS (1997) The GluR2 (GluR-B) hypothesis: Ca^{2+}-permeable AMPA receptors in neurological disorders. Trends Neurosci 20:464–470

Petito C, Feldmann E, Pulsinelli W, Plum F (1987) Delayed hippocampal damage in humans following cardiorespiratory arrest. Neurology 37:1281–1286

Pollard H, Héron A, Moreau J, Ben-Ari Y, Khrestchatisky M (1993) Alterations of the GluR-B AMPA receptor subunit flip/flop expression in kainate-induced epilepsy and ischemia. Neuroscience 57:545–554

Prince HK, Conn PJ, Blackstone CD, Huganir RL, Levey AI (1995) Down-regulation of AMPA receptor subunit GluR2 in amygdaloid kindling. J Neurochem 64:462–465

Pulsinelli WA, Brierley JB, Plum F (1982) Temporal profile of neuronal damage in a model of transient forebrain ischemia. Ann Neurol 11:491–498

Racca C, Catania MV, Monyer H, Sakmann B (1996) Expression of AMPA-glutamate receptor B subunit in rat hippocampal GABAergic neurons. Eur J Neurosci 8:1580–1590

Ribak CE, Nitsch R, Seress L (1990) Proportion of parvalbumin-positive basket cells in the GABAergic innervation of pyramidal and granule cells of the rat hippocampal formation. J Comp Neurol 300:449–461

Rump A, Sommer C, Gass P, Bele S, Meissner D, Kiessling M (1996) Editing of GluR2 RNA in the gerbil hippocampus after global cerebral ischemia. J Cereb Blood Flow Metab 16:1362–1365

Schmidt-Kastner R, Freund TF (1991) Selective vulnerability of the hippocampus in brain ischemia. Neuroscience 40:599–636

Schreiber SS, Baudry M (1995) Selective neuronal vulnerability in the hippocampus – a role for gene expression? Trends Neurosci 18:446–451

Seeburg PH (1993) The molecular biology of mammalian glutamate receptor channels. Trends Neurosci 16:359–365

Seeburg PH (1996) The role of RNA editing in controlling glutamate receeptor channel properties. J Neurochem 66:1–5

Sheardown MJ, Suzdak PD, Nordholm L (1993) AMPA, but not NMDA, receptor antagonism is neuroprotective in gerbil global ischemia, even when delayed 24 h. Eur J Pharmacol 236:347–353

Siesjö BK, Bengtsson F (1989) Calcium fluxes, calcium antagonists and calcium-mediated pathology in brain ischemia, hypoglycemia and spreading depression. A unifying hypothesis. J Cereb Blood Flow Metab 9:127–140

Silver IA, Erecinska M (1992) Ion homeostasis in rat brain in vivo: intra- and extracellular [Ca^{2+}] and [H^+] in the hippocampus during recovery from short- term, transient ischemia. J Cereb Blood Flow Metab 12:759–772

Sommer B, Keinänen K, Verdoorn TA, Wisden W, Burnashev N, Herb A, Köhler M, Takagi T, Sakmann B, Seeburg PH (1990) Flip and flop: a cell-specific functional switch in glutamate-operated channels of the CNS. Science 249:1580–1585

Sommer B, Köhler M, Sprengel R, Seeburg PH (1991) RNA editing in brain controls a determinant of ion flow in glutamate-gated channels. Cell 67:11–19

Stern-Bach Y, Bettler B, Hartley M, Sheppard PO, O'Hara PJ, Heinemann SF (1994) Agonist selectivity of glutamate receptors is specified by two domains structurally related to bacterial amino acid-binding proteins. Neuron 13:1345–1357

Szatkowski M, Attwell D (1994) Triggering and execution of neuronal death in brain ischemia: two phases of glutamate release by different mechanisms. Trends Neurosci 17:359–365

Tsubokawa H, Oguro K, Masuzawa T, Kawai N (1994) Ca^{2+}-dependent non-NMDA receptor-mediated synaptic currents in ischemic CA1 hippocampal neurons. J Neurophysiol 71:1190–1196

Tsubokawa H, Oguro K, Masuzawa T, Nakaima T, Kawai N (1995) Effects of a spider toxin and its analogue on glutamate-activated currents in the hippocampal CA1 neuron after ischemia. J Neurophysiol 74:218–225

Turetsky DM, Canzoniero LMT, Sensi SL, Weiss JH, Goldberg MP, Choi DW (1994) Cortical neurons exhibiting kainate-activated Co^{2+} uptake are selectively vulnerable to AMPA/kainate receeptor-mediated toxicity. Neurobiol Dis 1:101–110

Verdoorn TA, Burnashev N, Monyer H, Seeburg PH, Sakmann B (1991) Structural determinants of ion flow through recombinant glutamate receptor channels. Science 252:1715–1718

Virgo L, Samarasinghe S, de Belleroche J (1996) Analysis of AMPA receptor subunit mRNA expression in control and ALS spinal cord. Neuroreport 7:2507–2511

Weiss JH, Turetsky DM, Wilke G, Choi DW (1994) AMPA/kainate receptor-mediated damage to NADPH-diaphorase-containing neurons is Ca^{2+} dependent. Neurosci Lett 167:93–96

Wenthold RJ, Petralia RS, Blahos JI, Niedzielski AS (1996) Evidence for multiple AMPA receptor complexes in hippocampal CA1/CA2 neurons. J Neurosci 16:1982–1989

Wo ZG, Oswald RE (1995) Unraveling the modular design of glutamate-gated ion channels. Trends Neurosci 18:161–168

Ying HS, Weishaupt JH, Grabb M, Canzoniero LMT, Sensi SL, Sheline CT, Monyer H, Choi DW (1997) Sublethal oxygen-glucose deprivation alters hippocampal neuronal AMPA receptor expression and vulnerability to kainate-induced death. J Neurosci 17:9536–9544

Yoshioka A, Hardy M, Younkin DP, Grinspan JB, Stern JL, Pleasure D (1995) Alpha-amino-3-hydroxy-5-methyl-4-isoxazolepropionate (AMPA) receptors mediate excitotoxicity in the oligodendroglial lineage. J Neurochem 64:2442–2448

3 Factors Regulating Excitotoxic Neurodegeneration: The Role of Calcium and the Mitochondria

R. J. Miller, V. P. Bindokas, D. F. Babcock, J.-P. Lee, and J. B. Jordan

3.1 Introduction

Glutamate is the most widely used excitatory neurotransmitter in the central nervous system. In keeping with its important role in synaptic communication, we know that glutamate-mediated neurotransmission is mediated by a very large number of glutamate receptors (GluRs) which include families of both ionotropic (iGluRs) and metabotropic (mGluRs) receptors. Activation of these receptors can trigger several important phenomena. These include fast synaptic excitation, long-term potentiation, and excitotoxic neuronal degeneration. The molecular basis for these phenomena have been increasingly well understood over the last few years. For example, fast synaptic excitation depends on the influx of Na through activated ionotropic GluRs. On the other hand, longer-term glutamate-mediated events are linked to the influx of Ca through these receptor-gated ion channels (Choi 1988; Meldrum and Garthwaite 1990). Research from a number of laboratories has shown that the Ca permeability of ionotropic GluRs is subject to a number of control mechanisms which include the subunit composition of the receptor, editing of subunit specific mRNA, etc. Although the precise significance of all of these factors is not completely understood at this time, it appears that glutamate-activated Ca influx can potentially be very finely tuned (Burnashev et al. 1992).

There is no doubt about the importance of glutamate receptors in phenomena such as the induction of synaptic plasticity and neurodegeneration. Nevertheless, many key questions concerning the mechanism of action of glutamate remain to be answered. Let us consider the degeneration of neurons that is triggered by GluR activation – a process that is generally known as excitotoxicity (Rothman and Olney 1987). It is likely that this process is responsible for much of the neuronal death observed following stroke and epilepsy, for example. Moreover, there is increasingly compelling evidence that excitotoxicity is also important in neuronal death in many other diseases including Alzheimer's Disease, amyotrophic lateral sclerosis (ALS), Huntington's disease and AIDS dementia (Beal 1992, 1994, 1995; Charriaut-Marlangue et al. 1996; Lipton 1994; Porterra-Cailliau et al. 1995). It has also become increasingly clear that activation of Ca influx is probably the initial molecular event that is responsible for triggering excitotoxic neurodegeneration (Choi 1988). But what does this Ca actually do? How does it produce excitotoxicity when, under other circumstances, it produces synaptic plasticity and other important non-neurodegenerative phenomena? Furthermore, as all neurons possess GluRs, why is it that they exhibit differential sensitivity to the toxic effects of glutamate?

There have been plenty of suggestions as to why excessive levels of Ca might be neurotoxic and this is not surprising given the large number of cellular events that are potentially under the control of Ca. Moreover, many of these suggestions seem intuitively reasonable. Thus, it is easy to see why Ca-activated proteolysis or lipid degradation might lead to cell damage if left unchecked. However, although there is evidence supporting the participation of many such factors, no overall theory has found acceptance as an explanation for the molecular events underlying excitotoxicity.

One reason for this is that the situation is indeed complex. The concept of excitotoxicity was originally invoked to explain why neurons died in situations such as stroke. Our understanding as to the possible underlying molecular events involved was initially very rudimentary. However, our view of the mechanisms underlying cell death in general has now become very much more sophisticated. It is now no longer sufficient to talk of neuronal death, we must also consider what kind of death we are talking about. For example, in any particular circumstance we should be interested in whether death is necrotic, apoptotic, or whether elements of both of these processes are present (Bonfoco et al 1995; Charriaut-Marlangue et al. 1996). As a reflection of this complexity, opinions differ as to the major features involved in excitotoxic neurodegeneration. Studies can be found attesting to the fact that excitotoxicity is purely necrotic, purely apoptotic, or something in between (Bonfoco et al. 1995; Charriaut-Marlange et al. 1996; Porterra-Cailliau et al. 1997a,b; Zang and Geddes 1997). This is probably not a very useful argument. It is likely that excitotoxic stimulation of GluRs leads to a range of neuronal responses. Extremely intense activation and a very large Ca build up might rapidly activate destructive processes (e.g., proteolysis, free radical production, etc.) so that the cell is simply destroyed before it can mount any kind of longer-term response. However, if stimulation is less severe, longer-term events which require the activation of gene expression might also participate. Death may be more akin to apoptosis than necrosis under these circumstances (Bonfoco et al. 1995). Indeed, the notion of "slow excitotoxicity" may be particularly relevant in longer-term neurodegenerative diseases such as Alzheimer's disease, ALS, and Huntington's disease (Beal 1992, 1994, 1995).

In practice, therefore, it is not surprising that different types of excitotoxic death can be observed depending on the experimental paradigm employed. It is interesting to consider the potential role of Ca influx under all of these circumstances. For example, we have demonstrated that glutamate-mediated Ca influx can rapidly influence the function of neuronal mitochondria (Bindokas and Miller 1995). Waves of mitochondrial depolarization can be observed traversing neuronal circuits following synaptic activity in the hippocampus (Fig. 1). It is likely that signaling between synaptic activity and mitochondrial function is an essential element in the integration of cellular and synaptic activity. But how is this signaling achieved? Activation of neuronal

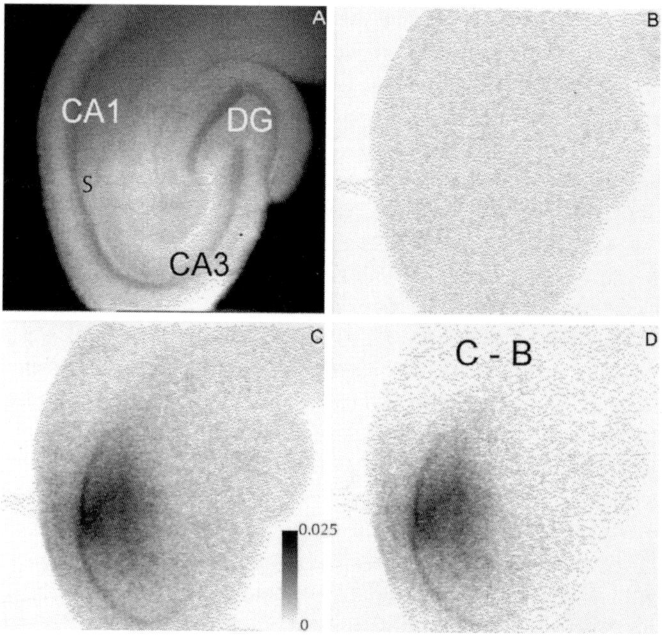

Fig. 1A–D. Electrical stimulation produces depolarization of mitochondria in the rat hippocampal slice. A thick slice was loaded with R123 and fluorescence changes in images (128 video-frame averages; approximately 5 s each) were analyzed by calculation of pixel-by-pixel variance in stacks of five images. **A** The digital average of the first five images of an image stack indicate the baseline staining pattern. The staining pattern is fairly uniform and the pyramidal layers appear dark. Indicated are the position of a bipolar stimulation electrode (*S*) and the approximate locations of the dentate gyrus (*DG*) and CA3 and CA1 regions. The width of the image is 3.3 mm. **B** The pixel-by-pixel standard deviation (SD) for five images averaged in (**A**) and shows little variability (range, 0–0.0062; mean SD=0.0015). **C** The SD for five images averaged during and after electrical stimulation trains of four stimuli (0.1 ms, 25 V, repeated at 100 Hz) repeated at 200-ms intervals for ten trains. Stimulation resulted in ictal-like bursts of action potentials recorded in the CA3 stratum pyramidale. Mitochondrial depolarization resulted in an increase in R123 fluorescence [here the maximum increase was 12 fluorescence units (FlU)] and therefore in an increase in SD (range, 0–0.0199; mean, 0.0032) over the approximate 25 s required to collect the image stack. This increase in SD is coded as the *darker gray* value. Note that most of the mitochondrial depolarization is confined to the strata pyramidale, oriens, and radiatum of the CA2, CA3 regions. **D** The digital difference between(**C**) and (**B**). The *gray scale* used in (**B**) and (**C**) corresponds to SDs ranging from 0–0.025 FIU

GluRs following addition of an agonist or the stimulation of excitatory synapses both produce increases in free intracellular Ca^{2+} concentration ($[Ca]_i$). Such $[Ca]_i$ increases are accompanied by mitochondrial depolarization as a result of Ca uptake into these organelles. Indeed, Ca is taken up by mitochondria in response to cellular Ca signals in both the physiological and pathological range (Miller 1991). The consequences of this Ca uptake are likely to be extremely important for numerous reasons. For example, it is likely that this is a way of coupling the activity of neurons to their energetic requirements through the regulation of Ca-sensitive components of the tricarboxylic acid cycle (TCA) cycle located in the mitochondrial matrix (McCormack et al. 1990). In addition, storage of Ca by mitochondria and subsequent Ca buffering by these organelles have been suggested as being important elements in certain types of synaptic plasticity (Miller 1991; Herrington et al. 1996; Tang and Zucker 1997).

Neuronal mitochondria are ideally equipped to cooperate with Ca in the sensing and integrating of many aspects of cellular activity and then coordinating the appropriate response to these events. This is certainly true with respect to both necrotic and apoptotic aspects of excitotoxic neurodegeneration. For example, mitochondrial Ca uptake can trigger the production of superoxide (O_2^-) and other free radicals that may be important mediators of both short and long-term neuronal events (Fig. 2) (Bindokas et al. 1996). As will be further discussed below, O_2^- may give rise to other potentially destructive reactive oxygen species (ROS). For example, O_2^- produced by the mitochondria can react with NO to produce the potentially toxic peroxynitrite radical (Lipton et al. 1993). In addition, O_2^- and other ROS may also lead to the induction of genes, such as the p53 tumor suppressor gene, which have been shown to be important mediators of apoptosis in many cell types including neurons (Jordan et al. 1997). Furthermore, Ca uptake by the mitochondria can lead to substantial changes in the permeability of the inner mitochondrial membrane through opening of the mitochondrial permeability transition pore (MPT). The MPT is a very large channel that is permeable to molecules up to 1500 kDa (Scorrano et al. 1997). Following activation of the MPT, accumulated Ca can leave the mitochondrial matrix rapidly. This can be looked upon as a way that the mitochondria can rapidly dump accumulated Ca. Long-term opening of the MPT may lead to a decline in mitochondrial potential, in the synthesis of ATP, and

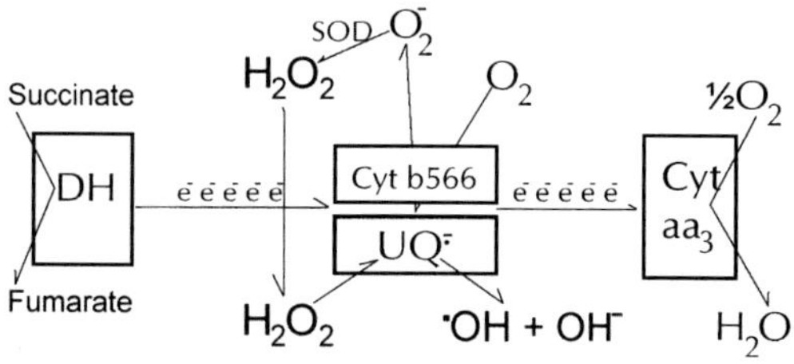

Fig. 2. Sources of mitochondrial free radical production. *SOD*, superoxide dismutase; *DH*, dehydrogenase; *UQ*, ubiquinone

subsequent triggering of neuronal death. Indeed, it has been shown that cyclosporin A, a drug that effectively blocks the MPT, can inhibit excitotoxic death in some paradigms (Schinder et al. 1996; Uchino et al. 1995). However, the transient opening and closing of the MPT may also have an important role in Ca signaling.

Further insights into the potential key role of the mitochondria has come from recent studies on the basic mechanisms underlying apoptosis. It has been shown that key elements in the final pathway involved in the execution phase of apoptosis in most cells, including Ced3/caspase, Ced4 and its mammalian homologue, and Ced9/bcl-2 exist in a complex that is tethered to bcl-2 localized in the outer mitochondrial membrane (Golstein 1997). Many functions have been suggested for bcl-2 and related antiapoptotic proteins, including interaction with Ca-sensing elements such as calcineurin (Shibasaki et al. 1997). It has been demonstrated that changes in mitochondrial function occur during the execution of apoptosis and that proteins such as bcl-2 can stabilize mitochondrial function during these events (Marchetti et al. 1996). For example, expression of bcl-2 opposes the opening of the MPT. In summary, the mitochondria are clearly of central importance for the integration of a great deal of activity connected with the death of neurons in several circumstances.

A further important phenomenon to consider is that of selective vulnerability. All neurons possess GluRs, but not all neurons are equally sensitive to excitotoxic stimulation. Indeed, selective populations of neurons die in all neurodegenerative diseases (Beal 1992, 1994, 1995). In familial ALS for example, every cell expresses mutant superoxide dismutase (SOD-1), yet motor neurons are selectively targeted. It is now clear that there are an enormous number of risk factors that influence the susceptibility of particular neurons to particular insults. In the case of excitotoxicity, these factors may include the presence of Ca-binding proteins (e.g., calbindin D_{28k}) (Mattson et al. 1991), or enzymes that are important in the removal of ROS [e.g., superoxide dismutase (SOD) and nitric oxide synthase (NOS)] (Bindokas et al. 1996; Lipton et al. 1993). A further observation of great interest concerns the results of activating mGluRs. It has been clearly demonstrated that activation of different classes of mGluRs may either enhance or ameliorate excitotoxic damage – although the molecular mechanisms by which this occurs are unknown (Nicoletti et al. 1996). As the population of mGluRs may differ on different neurons, coactivation of these receptors, together with iGluRs, may engender some specificity to the excitotoxic response.

In a broader context one should consider that in a number of neurodegenerative disorders, where apoptotic death appears to be of major importance, the initial triggers for death may differ even if the ultimate events in the execution phase are the same. Thus, although the final common pathway for the execution of apoptosis may be common to every neuron or even every cell, the triggering mechanisms may be cell-specific. Hence, there are a considerable number of ways in which neurons could become selectively vulnerable. Perhaps it would be useful to look at excitotoxicity the other way round and postulate that excessive stimulation of GluRs on all neurons would lead to death unless they are selectively saved by one or more modulatory factors. What are these modulatory factors and how do they work? In this chapter, we shall further consider several of the possibilities.

3.2 Regulation of Intracellular Ca and Mitochondrial Function

As we have discussed above, the activation of GluRs produces both short- and long-term changes that are connected with the induction of both apoptotic and necrotic neuronal death. Many of these changes are triggered by increases in neuronal $[Ca]_i$. Activation of GluRs can lead to increases in neuronal $[Ca]_i$ by several mechanisms. The most obvious of these is by direct Ca influx via Ca-permeable ionotropic GluRs. In addition, depolarization produced by activation of iGluRs can activate

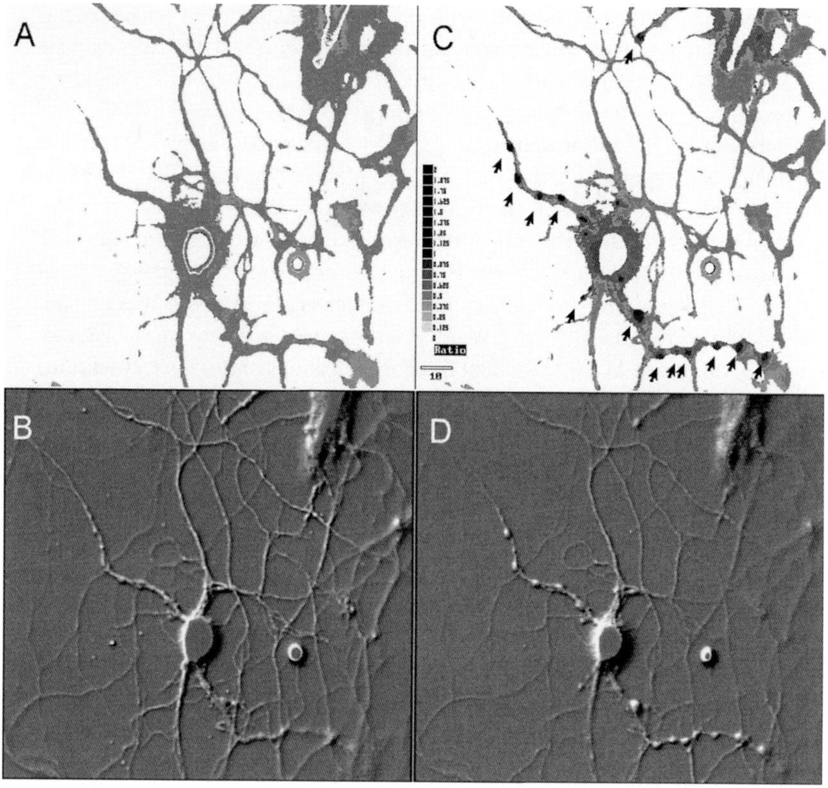

Fig. 3. Blebs tend to only form in neurites in cultured rat cerebellar neurons exhibiting a large rise in free intracellular Ca^{2+} concentration ($[Ca^{2+}]_i$). **A** Intensity-coded image of fura-2 fluorescence (f340/f380) under basal conditions. **B** Control image derived from fura-2 emission (Ex 380 nm) by digital filtering. **C** f340/f380 Ratio image taken 1 min after initiation of kainic acid (KA) application (200 μM) shows $[Ca^{2+}]_i$ rise was restricted to only a few neurites and that the rise in $[Ca^{2+}]_i$ was also nonuniform within those neurites. The regions with the greatest increases in $[Ca^{2+}]_i$ (higher f340/f380 ratio) occurred primarily in two neurites of the cell shown at the *center* of the image, as well as in a neurite at the *top* of the field. Signals from the soma regions of this neuron and glia at the *top* of the field saturated the camera and appear *black* in this image. **B**, **D** Blebs formed (8 min after start of KA) only on the neurites with large rises in $[Ca^{2+}]_i$. Note the blebs in the neurite at the *top* of the field; these blebs are also associated with neurite intersections, and with regions that exhibited the highest $[Ca^{2+}]_i$. Peaks in the $[Ca^{2+}]_i$ accurately predicted the sites at which blebs would later form

Ca influx through voltage-sensitive Ca channels. Furthermore, activation of mGluRs can produce both Ca mobilization from intracellular stores and also influence Ca influx through iGluRs by a variety of mechanisms.

Given the fact that iGluRs tend to be clustered at synaptic sites, it is reasonable that the activation of iGluRs produces Ca influx in a heterogeneous fashion (Fig. 3) (Bindokas and Miller 1995). The increases in $[Ca]_i$ in these synaptic regions lead to a breakdown in local ionic gradients and to local swellings or bleb formation. It is thought that these blebs may be the forerunners of the degeneration of neurites. It is likely that this kind of mechanism is an important one by which neuronal damage can be produced without the participation of gene transcription. Presumably, such mechanisms make an important contribution to rapid excitotoxic necrosis.

However, there are many other important consequences of glutamate-activated neuronal $[Ca]_i$ increases. As discussed above, one of the most important appears to be the modulation of mitochondrial function. Thus, it is clear that following iGluR activation, much of the Ca flowing into the cell can enter the mitochondria (Bindokas et al. 1996; White and Reynolds 1997). The result of this Ca entry is the collapse of the mitochondrial membrane potential, an event that can be observed using the fluorescent dye rhodamine 123. This dye enters the mitochondrial matrix under the influence of the mitochondrial membrane potential (approximately –150 mV), and its fluorescence is quenched. Upon collapse of the membrane potential, the dye can leave the mitochondrial matrix and an increase in fluorescence is observed. It is clear that the mitochondrial depolarization induced by activation of GluRs is due to the influx of Ca rather than Na, as it is not observed in Ca free medium (Bindokas and Miller 1995).

The effect of GluR activation on mitochondrial function can be observed in another context in slices of the hippocampus. Epileptiform synaptic activity [stimulation-induced bursting (STIB)] can be induced by stimulation with an electrode in the stratum lucidum of the CA3 in Mg-free medium. Following the induction of STIB, waves of mitochondrial depolarization can be seen traversing the slice along defined neuronal circuits (see Fig. 1). Manipulations that are designed to inhibit GluR function (i.e., the GluR blockers CNQX/APV) or to reduce the release of glutamate by a presynaptic mechanism [e.g., tetrodotoxin (TTX) or by activation of presynaptic neuropeptide Y (NPY) receptors] reduce both synaptic bursting and the intensity of mitochondrial waves. Mitochondrial waves often proceed in a saltatory fashion, pausing at a particular point before continuing on their way.

It is clear that the time course of depolarizing mitochondrial waves is much slower than that of the accompanying synaptic activity. This raises the question as to the mechanism of propagation of these waves. As we have discussed above, the collapse of the mitochondrial membrane potential is due to the influx of Ca. Presumably, the movement of Ca also underlies the propagation of the waves. Following the entry of Ca into the mitochondrial matrix, the efflux of Ca can occur by at least two mechanisms. The first is by exchange for another cation, which in the case of neurons is usually Na. The efflux of stored mitochondrial Ca in this way is responsible for the buffering of the cytoplasmic Ca at the level of the mitochondrial set-point (Nicholls and Akerman 1982; Thayer and Miller 1990). This phenomenon, which has been frequently observed, may underlie some aspects of synaptic plasticity (Miller 1990; Tang and Zucker 1997). A second mechanism might involve the opening of the MPT (Scorrano et al. 1997; Marchetti et al. 1996). As discussed above, this mega-channel opens in the inner mitochondrial membrane in response to loading with Ca, and it is also catalyzed by oxidizing agents such as reactive free radicals (ROS, see below) and depolarization. The MPT is a very large channel that is permeable to species of MW up to 1500 daltons. One result of the opening of the MPT is the immediate dumping of Ca that had been accumulated by the mitochondrion. It has been generally thought that the opening of the MPT and accompanying depolarization is generally irreversible and heralds the death of the cell. Indeed, drugs that can block the MPT can ameliorate excitotoxic neuronal death (Schinder et al. 1996). However, it is also possible that reversible opening and closing of the MPT might be a way of allowing waves of Ca to traverse the cell accompanied by waves of mitochondrial depolarization. In keeping with this possibility, preliminary results from our laboratory show that cyclosporin A can reduce the amplitude of mitochondrial waves in the hippocampal slice.

3.3 Free Radical Generation

What is the significance of signaling to the mitochondria as the result of the activation of GluRs? As discussed above, it is likely that the mitochondria may be central integrators of the function of the neuron in several respects. Entry of Ca into the mitochondrial matrix is of great importance in the control of Ca-sensitive enzymes of the TCA cycle and is therefore a way of coupling synaptic activity to the energetic requirements of the cell. Another consequence of mitochondrial Ca uptake, however, is that the stimulation of respiration results in increased production of superoxide free radicals (O_2^-) (see Fig. 2) (Bindokas et al. 1996; Reynolds and Hastings 1995). Mitochondria may also generate other ROS such as OH·. The production of ROS may have many consequences for both short- and long-term neuronal regulation. In particular, the reaction of O_2^- with NO results in the formation of the highly reactive peroxynitrite ion, which can nitrosylate proteins, thereby altering their function (Lipton et al. 1993). Indeed, owing to the fact that NOS, the enzyme responsible for NO synthesis, is also activated by Ca, the formation of peroxynitrite will be favored under conditions of stimulated Ca influx (Lipton et al. 1993). Furthermore, ROS are important regulators of gene transcription, primarily through effects such as the activation of transcription factors such as nuclear factor (NF)-κB (Togashi et al. 1997). Thus, the production of ROS could be another way in which activation of GluRs could produce both short- and long-term regulation of neuronal function.

It is clear that activation of iGluRs can activate O_2^- production in neurons (Bindokas et al. 1996; Reynolds and Hastings 1995). This can be observed in Figs. 4 and 5, in experiments on cultured hippocampal pyramidal neurons. Sites of active O_2^- production can be seen to correspond closely with mitochondria and are dependent on the influx of Ca through iGluRs. The production of O_2^- is enhanced by mitochondrial uncouplers such as trifluoromethoxycarbonyl-cyanide-phenylhydrazone (FCCP) and inhibited by cyanide (CN^-), also indicating that the mitochondria are the major site of production. It is interesting to note that addition of NO donors such as SNOC, reduces the appearance of O_2^-, presumably due to the formation of peroxynitrite. On the other hand, there is a population of cells in the hippocampal neuronal cultures whose O_2^- production is enhanced by NOS blockers such as N^G-nitro-L-

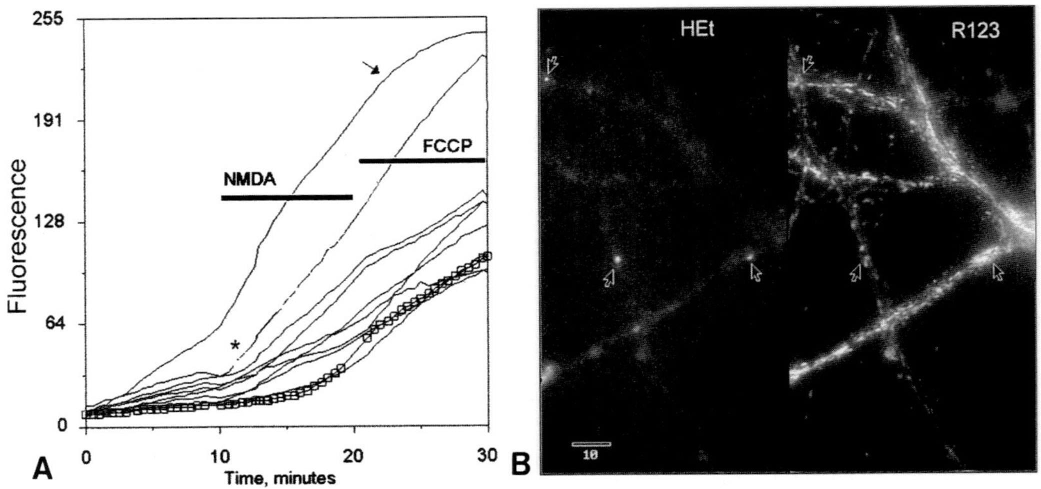

Fig. 4A,B. Continuous records of O_2^- production in cultured rat hippocampal pyramidal neurons using hydroethidine (*HEt*) for detecting O_2^- (Bindokas et al. 1996). **A** Results from a group of cultured neurons illustrating the stimulatory effects of *N*-methyl-D-aspartate (*NMDA*). The increase in O_2^- production was delayed to varying degrees ranging from nearly immediate (*curve* marked by asterisk) to 7 min after application (*open squares*). O_2^- production, on average, was not increased additionally by trifluoromethoxycarbonyl-cyanide-phenylhydrazone (*FCCP*). This implies that mitochondria were fully depolarized by the NMDA treatment. The signal from one neuron approaches camera saturation (*arrow*). **B** High-magnification images of initial HEt staining pattern revealed distinct punctata (*arrows*). The *left image* was taken approximately 3 min after application of HEt solution, while the *right image* pair was taken after 2 min staining with 10 μg/ml (26 nM) R123 and at lower camera gain. Numerous brightly stained mitochondria are evident, and some were colocalized with HEt-stained regions (*arrows*). Thus, all HEt-stained regions colocalized with mitochondria, but the converse was not true. Scale bar, 10 μM arginine methylester (L-NAME), suggesting that they are actively producing NO, presumably in response to GluR-activated Ca influx (Fig. 5). In addition to the situation in cell culture, increased excitatory synaptic activity in hippocampal slices also produces increases in O_2^- production, particularly in the presence of L-NAME to suppress endogenous NO production.

The potential consequences of increased NO production and disturbances of ROS production for neuronal viability can be seen in studies on the overexpression of endothelial cell nitric oxide synthases (eNOS) in hippocampal neurons with an adenovirus. Preliminary data from our laboratory indicates that overexpression of eNOS increases the toxic effects of *N*-methyl-D-aspartate (NMDA) application, whereas overexpression of WTSOD-1 or the Ca-binding protein, calbindin D_{28k}, reduces it.

3.4 The Consequences of mGluR Activation

In addition to iGluRs, it has also been shown that activation of mGluRs can have an impact on neuronal viability (Nicoletti et al. 1996). mGluRs can be activated by exogenous agonists, but also by synaptically released glutamate, particularly during intense synaptic activity (Glaum and Miller 1993). Indeed, activation of mGluRs by glutamate released following a tetanus seems to have profound effects on the establishment of some forms of synaptic plasticity (Bashir et al. 1993). At least three families comprising eight separate members of mGluRs have been identified to date (Pin and Duvoisin 1995). The precise consequences of mGluR activation for neuronal viability depend on the subtype(s) of mGluR activated (Nicoletti et al. 1996; Pisani et al. 1997). However, it has frequently been shown that activation of type 2 and 3 mGluRs can reduce excitotoxic damage. How does activation of class 2/3 mGluRs reduce excitotoxicity? Although the answer to this question is not yet completely clear, it is likely that there are both pre- and postsynaptic sites of action. Activation of mGluRs in hippocampal neuronal cultures depresses NMDA-induced Ca signals. mGluR activation reduces a component of the NMDA-induced Ca signal even after TTX treatment, although this component is reduced in magnitude. It is interesting to note that TTX also ameliorates NMDA-induced toxicity. Thus, it ap-

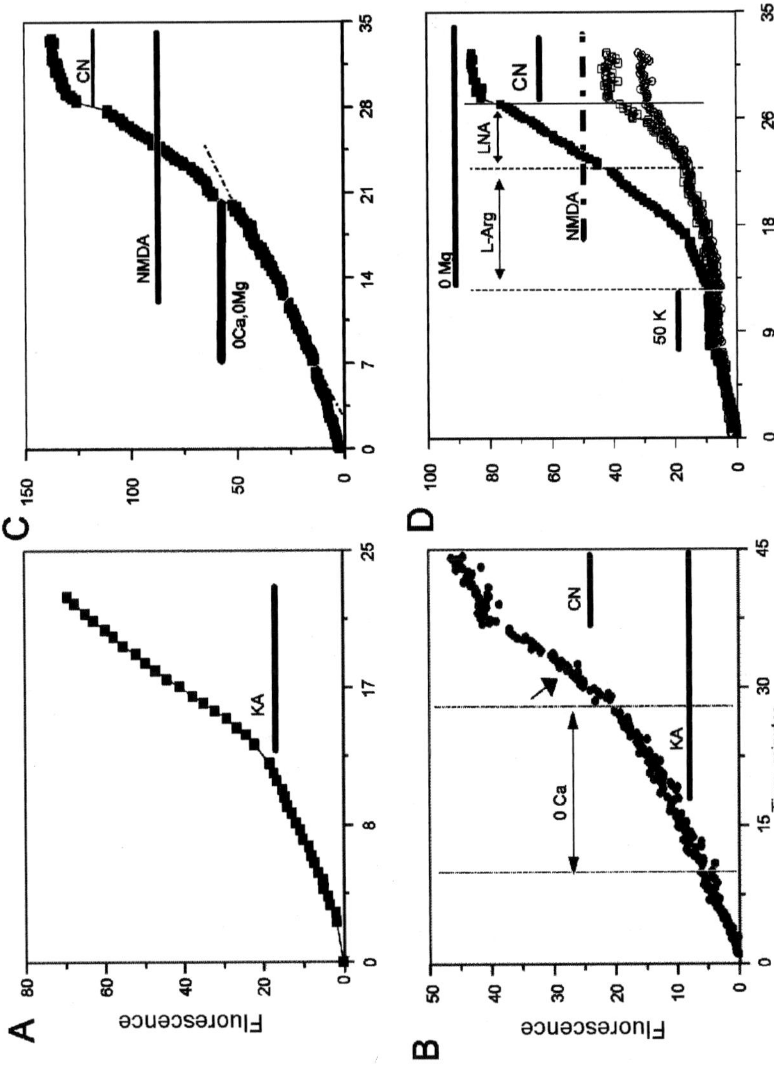
A
B
C
D
Fluorescence
Time, minutes
KA
0 Ca
CN
NMDA
0Ca,0Mg
0 Mg
L-Arg
LNA
50 K

pears that activation of mGluRs may inhibit the effects of NMDA both by reducing NMDA-induced release of glutamate and also by inhibiting NMDA-induced Ca influx through some postsynaptic mechanism.

There is good evidence to support both of these mechanisms. It has been known for many years that some mGluRs have a presynaptic localization. Stimulation of these receptors can produce presynaptic inhibition of evoked transmitter release. These effects may be a result of the observed ability of mGluRs to inhibit the voltage-sensitive Ca channels that are responsible for triggering transmitter release (Glaum and Miller 1993). In addition, it is also clear that the localization of some mGluRs is postsynaptic. Activation of these receptors can have a variety of consequences, including the modulation of different conductances, including K channels and also NMDA receptors (Glaum and Miller 1992). In the latter case, both enhancement and inhibition of NMDA currents have been reported depending on the paradigm employed (Bleakman et al. 1992; Yu et al. 1997).

3.5 Long-Term Consequences of Ca Influx

The interactions between Ca and mitochondrial function also provide a framework for considering some of the long-term influences on excitotoxic neurodegeneration. As we have discussed above, activation of GluRs may produce neuronal death by mechanisms that include both necrotic and apoptotic events (Porterra-Cailliau et al. 1997a,b). It is well established that many factors can modulate apoptotic death. For exam-

◀ **Fig. 5A–D.** O_2^- production in cultured rat hippocampal pyramidal neurons detected using hydroethidine (HEt) (Bindokas and Miller 1995). **A** Stimulation with kainic acid (*KA*). **B** KA stimulation was dependent on external Ca and blocked by cyanide (CN^-). **C** *N*-methyl-D-aspartate (*NMDA*) stimulation was also Ca-dependent and blocked by CN. **D** O_2^- levels can be influenced by nitric oxide synthase (NOS) activity. Neurons were washed immediately with Mg^{2+}-free saline (*0 Mg*) containing L-arginine (*L-Arg*, 3 mM) and remained in Mg^{2+}-free saline for the remainder of the experiment. NMDA (100 µM) + L-Arg increased the rate of HEt oxidation, but the rate remained constant in two neurons (*open squares*). Replacement of L-Arg with the NOS inhibitor linolenic acid (*LNA*) (100 µM) resulted in an increase in slope in both of these neurons. Finally, CN^- was able to block HEt oxidation in all cases

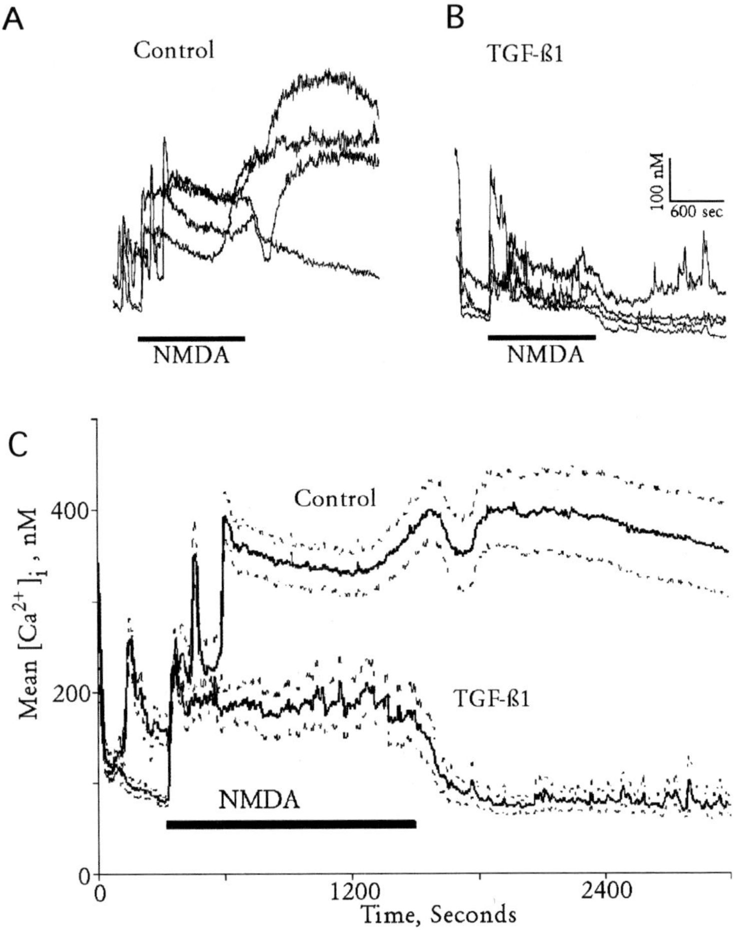

Fig. 6A–C. Protective effects of transforming growth factor (TGF)-β1 on cultured hippocampal pyramidal neurons. **A** Baseline fluctuations in free intracellular Ca^{2+} concentration ($[Ca^{2+}]_i$) represent ongoing network activity at excitatory synapses established between cultured neurons. Four representative neurons are shown. Application of *N*-methyl-D-aspartate (NMDA) (100 μM; 20 min) produced sustained elevations in $[Ca^{2+}]_i$ in most neurons that outlasted application. Note secondary increases in three or four neurons shown. **B** A 2-h pretreatment with 10 ng of TGF-β1/ml reduced peak $[Ca^{2+}]_i$ and plateau level during NMDA application. TGF-β1 was present during NMDA exposure. Rapid recovery was evident in all four representative neurons. **C** Average data from 22 cells ple, following GluR activation, the product of the p53 tumor suppressor gene may be of significance (Morrison et al. 1996). Expression of this protein in dividing cells has two major effects: inhibition of cell division and the induction of apoptosis. In addition, it has been shown that treatment of neurons with GluR agonists or ischemic conditions can cause the induction of p53 in neurons (Sakhi et al. 1994). However, the question remains as to the consequences of inducing p53 in non-dividing cells such as neurons. Given the fact that p53 has been shown to be induced in many cell types by X-irradiation of the type that produces DNA damage, we examined the effects of X-irradiation hippocampal pyramidal neurons in culture (Jordan et al. 1997). We observed that X-irradiation killed neurons by apoptosis and that this was accompanied by the induction of p53. In addition, when we induced p53 synthesis in these neurons using a p53-containing adenovirus, they also died by apoptosis. Interestingly, X-irradiation-induced neuronal death could be blocked by expression of several other gene products including the retinoblastoma gene product. These results illustrate the types of events that may occur downstream of Ca entry in the induction of neuronal apoptosis as a result of GluR activation.

Other gene products can impact cell death by opposing apoptosis. Included are proteins such as bcl-2 and other members of this family. The effects of bcl-2 further highlight the role of the mitochondria in the central control of cell viability. It has been shown that bcl-2 is localized primarily to the outer mitochondrial membrane, where it may form a complex with other key proteins that carry out the final execution phase of apoptosis – including ced-3/caspase and ced-4 and its mammalian homologue (Golstein 1997). It is interesting to note that among other things, the presence of bcl-2 seems to stabilize the mitochondrial membrane potential in the face of various cellular stresses, so that the mitochondria can take up more Ca. Expression of bcl-2 has been shown to prevent the opening of the MPT in several instances (Marchetti et al. 1996). An example of these effects is provided by our studies on the mechanism of action of the cytokine transforming growth factor (TGF)-β1 (Prehn et al. 1995). We showed that this cytokine, which is neuroprotective in stroke models in vivo, was also able to protect neurons from the effects of NMDA in culture. In association with its neuroprotective effects, we also observed that treatment with TGF-β1 was associated with upregulation of both bcl-2 and bcl-xl, with no effect on bax levels.

The effects of TGF-β1 treatment on neuronal Ca were also most interesting. Normally, long-term treatment with NMDA produced large increases in Ca that reached a plateau that outlasted the application of NMDA. This plateau is probably due to buffering of the cytoplasmic Ca as it leaves the mitochondria. Continued treatment with NMDA finally results in loss of the cell's ability to regulate Ca whose levels then rise precipitously. This may be due to activation of the MPT. As can be observed in Fig. 6, treatment with TGF-β1 resulted in stabilization of the mitochondrial membrane potential and more efficient regulation of Ca levels. The improvement in Ca regulation occurred not only following treatment with NMDA, but also with a Ca ionophore such as ionomycin, indicating that the effects of TGF-β1 were probably exerted downstream of Ca entry via the NMDA receptor.

3.6 Conclusions

It is clear that the activation of GluRs can lead to a large number of cellular consequences, many of which can effect the viability of neurons. Many of these events are the result of increased Ca influx resulting from receptor stimulation. As pointed out in this article and elsewhere, it is likely that the mitochondria are important targets for the Ca that is taken up by neurons. Regulation of mitochondrial function by Ca, including the generation of free radicals, appears to be an important mechanism for integrating synaptic activity and the energetic and other requirements of the neuron.

References

Bashir ZI, Bortolotto ZA, Davies CH, Berretta N, Irving AJ, Seal AJ, Henley JM, Jane DE, Watkins JC, Collingridge GL (1993) Induction of LTP in the hippocampus needs synaptic activation of glutamate metabotropic receptors. Nature 363:347–350

Beal MF (1992) Does impairment of energy metabolism result in excitotoxic neuronal death in neurodegenerative illnesses. Ann Neurol 31:119–130

Beal MF (1994) Neurochemistry and toxin models in Huntington's disease. Curr Opinion Neurol 7:542–547

Beal MF (1995) Aging, energy and oxidative stress in neurodegenerative disease. Ann Neurol 38:357–366

Bindokas VP, Miller RJ (1995) Excitotoxic neurdegeneration is initiated at non random sites in cultured rat cerebellar neurons. J Neurosci 15:6999–7011

Bindokas VP, Jordan J, Lee C, Miller RJ (1996) Superoxide production in rat hippocampal neurons; selective imaging with hydroethidine. J Neurosci 16:1324–1336

Bleakman D, Rusin KI, Chard PS, Glaum SR, Miller RJ (1992) Metabotropic glutamate receptors potentiate ionotropic glutamate responses in the rat dorsal horn. Mol Pharmacol 42:192–196

Bonfoco E, Krain C, Ankarcrona M, Nicotera P, Lipton SA (1995) Apoptosis and necrosis; two distinct events induced, respectively, by mild and intense insults with N-methyl-D-aspartate or nitric oxide/superoxide in cortical cell cultures. Proc Natl Acad Sci USA 92:7162–7166

Burnashev N, Monyer H, Seeburg PH, Sakmann B (1992) Divalent ion permeability of AMPA receptor channels is dominated by the edited form of a single subunit. Neuron 8:189–198

Charriaut-Marlangue C, Aggoun-Zouaoui D, Represa A, Ben-Ari Y (1996) Apoptotic features of selective neuronal death in ischemia, epilepsy and gp120 toxicity. Trends Neurosci 19:109–114

Choi DW (1988) Calcium mediated neurotoxicity: relationship to specific channel types and role in ischemic damage. Trends Neurosci 11:465–469

Glaum SR, Miller RJ (1992) Metabotropic glutamate receptors mediate excitatory transmission in the nucleus of the solitary tract. J Neurosci 12:2251–2259

Glaum SR, Miller RJ (1993) Metabotropic glutamate receptors depress afferent excitatory transmission in the rat nucleus tractus solitarius. J Neurophysiol 70:2669–2672

Golstein P (1997) Controlling cell death. Science 275:1081–1082

Herrington J, Park YB, Babcock DF, Hille B (1996) Dominant role of mitochondria in clearance of large Ca loads from rat adrenal chromaffin cells. Neuron 16:219–228

Jordan J, Galindo MF, Prehn JHM, Weichselbaum RR, Beckett M, Ghadge GD, Roos RP, Leiden JM, Miller RJ (1997) p53 expression induces apoptosis in hippocampal pyramidal neuronal cultures. J Neurosci 17:1397–1405

Lipton SA (1994) Ca, NMDA receptors and AIDS related neuronal injury. Int Rev Neurobiol 36:1–27

Lipton SA, Choi YB, Pan ZH, Lei SZ, Chen HS, Sucher NJ, Loscalzo J, Singel DJ, Stamler JS (1993) A redox based mechanism for the neuroprotective and neurodestructive effects of nitric oxide and related nitroso compounds. Nature 364:626–632

Marchetti P, Castedo M, Susin SA, Zamzami N, Hirsch T, Macho A, Haeffner A, Hirsch F, Geuskens M, Kroemer G (1996) Mitochondrial permeability transition is a central coordinating event of apoptosis. J Exp Med 184:1155–1160

Mattson MP, Rychlik GC, Chu C, Christakos S (1991) Evidence for Ca reducing and excitoprotective roles for Ca binding protein Calbindin O_{28K} in cultured hippocampal neurons. Neuron 68:41–51

McCormack JG, Halestrap AP, Denton RM (1990) Role of calcium ions in regulation of mammalian intramitochondrial metabolism. Physiol Rev 70:391–425

Meldrum B, Garthwaite J (1990) Excitatory amino acid neurotoxicity and neurodegenerative disease. Trends Pharmacol Sci 11:379–387

Miller RJ (1991) The control of neuronal Ca homeostasis. Prog Neurobiol 37:255–285

Morrison RS, Wenzel HJ, Kinoshita Y, Robbins CA, Donehower LA, Schwartzkroin PA (1996) Loss of the p53 tumor suppressor gene protects neurons from kainate induced cell death. J Neurosci 16:1337–1345

Nicholls DG, Akerman KEO (1982) Mitochondrial Ca transport. Biochem Biophys Acta 683:57–88

Nicoletti F, Bruno V, Copani A, Casabona G, Knopfel T (1996) Metabotropic glutamate receptors; a new target for the therapy of neurodegenerative disorders? Trends Neurosci 19:267–271

Pin JP, Duvoisin R (1995) Neurotransmitter receptors 1. The metabotropic glutamate receptors; structure and functions. Neuropharmacology 34:1–26

Pisani A, Calabresi P, Centonze D, Benardi G (1997) Enhancement of NMDA responses by group 1 metabotropic glutamate receptor activation in striatal neurons. Br J Pharmacol 120:1007–1014

Porterra-Cailliau C, Hedreen JC, Price DL, Koliatsos VE (1995) Evidence for apoptotic death in Huntington's disease and excitotoxic animal models. J Neurosci 15:3775–3787

Porterra-Cailliau C, Price DL, Martin LJ (1997a) Excitotoxic neuronal death in immature brain is an apoptosis – necrosis morphological continuum. J Comp Neurol 378:70–87

Porterra-Cailliau C, Price DL, Martin LJ (1997b) Non NMDA and NMDA receptor mediated excitotoxic neuronal deaths in adult brain are morphologically distinct; further evidence for an apoptosis – necrosis continuum. J Comp Neurol 378:88–104

Prehn JHM, Bindokas VP, Marcuccilli CJ, Krajewski S, Reed JC, Miller RJ (1995) Regulation of neuronal bcl-2 expression and calcium homeostasis by transforming growth factor b1 confers wide ranging protection on rat hippocampal neurons. Proc Natl Acad Sci USA 91:12599–12603

Reynolds IJ, Hastings TG (1995) Glutamate induces the production of reactive oxygen species in cultured forebrain neurons following NMDA receptor activation. J Neurosci 15:3318–3327

Rothman SM, Olney JW (1987) Excitotoxicity and the NMDA receptor. Trends Neurosci 10:299–302

Sakhi S, Bruce A, Sun N, Tocco G, Baudry M, Schreiber SS (1994) P53 induction is associated with neuronal damage in the central nervous system. Proc Natl Acad Sci USA 91:7525–7529

Schinder AF, Olson EC, Spitzer NC, Montal M (1996) Mitochondrial dysfunction is a primary event in glutamate neurotoxicity. J Neurosci 16:6125–6133

Scorrano L, Petronelli V, Bernardi P (1997) On the voltage dependence of the mitochondrial permeability transition pore. J Biol Chem 272:12295–12999

Shibasaki F, Kondon E, Akagi T, McKeon F (1997) Suppression of signalling through transcription factor NF-AT by interactions between calcineurin and Bcl-2. Nature 386:728–731

Tang YG, Zucker RS (1997) Mitochondrial involvement in post tetanic potentiation of synaptic transmission. Neuron 18:483–491

Thayer SA, Miller RJ (1990) Regulation of intracellular free calcium concentration in single rat dorsal root ganglion neurons in vitro. J Physiol (Lond) 425:85–115

Togashi H, Sasaki M, Frohman E, Taira E, Ratan RR, Dawson TM, Dawson VL (1997) Neuronal (type1) nitric oxide synthase regulates nuclear factor KB activity and immunologic (type 2) nitric oxide synthase expression. Proc Natl Acad Sci USA 94:2676–2680

Uchino H, Elmer E, Uchino K, Lindvall O, Siesjö BK (1995) Cyclosporin A dramatically ameliorates CA1 hippocampal damage following transient forebrain ischaemia in the rat. Acta Physiol Scand 155:469–471

White RJ, Reynolds IJ (1997) Mitochondria accumulate Ca following intense glutamate stimulation of cultured rat forebrain neurons. J Physiol (Lond) 498:31–47

Yu SP, Sensi SC, Canzoniero LMT, Buisson A, Choi DW (1997) Membrane delinated modulation of NMDA currents by metabotropic glutamate receptor subtypes 1/5 in cultured mouse cortical neurons. J Physiol (Lond) 499:721–732

Zang P, Geddes JW (1997) Mechanisms of cell death induced by the mitochondrial toxin 3-nitroproprionic acid: acute excitotoxic necrosis and delayed apoptosis. J Neurosci 17:3064–3073

4 Excitotoxicity, Genetics and Neurodegeneration in Amyotrophic Lateral Sclerosis

P.J. Shaw

4.1 Introduction

Amyotrophic lateral sclerosis (ALS) is one of the commonest human neurodegenerative disorders of adult life. The cell-death process in ALS is relatively selective for upper motor neurones, a proportion of which are represented by Betz cells (Fig. 1A) in the fifth layer of the motor cortex, and for lower motor neurones in the ventral horn of the spinal cord (Fig. 1B) and brainstem. The selective vulnerability of motor neurones in ALS is relative and, there is now increasing evidence from pathological (Ince et al. 1996; Iwanaga et al. 1997; Williams et al. 1995) and clinical (Chari et al. 1996; Kew et al. 1993; Subramaniam and Yiannikas 1990) studies that extra-motor system involvement commonly occurs, and that ALS is in fact a multisystem disorder. As a result of the neurodegenerative process, the individual afflicted by ALS develops progressive muscle weakness, wasting and spasticity and usually dies from respiratory failure within 3–5 years from the onset of symptoms. One of the interesting clinical features in ALS is that certain groups of motor neurones tend to be selectively spared from injury, including those brainstem nuclei innervating extraocular muscles responsible for eye movement and those in Onuf's nucleus in the sacral spine cord innervating the muscles of the pelvic floor.

The degenerating motor neurones in ALS are characterised by the presence of ubiquitinated inclusion bodies (Leigh et al. 1988; Lowe et al. 1988). The inclusions may be hyaline, skein-like or Lewy body-like (Fig. 2A,B). In most circumstances, the major protein constituent of the fibrils comprising these inclusions is unknown although the most promising candidates are neurofilament epitopes. In some cases of familial ALS, hyaline conglomerate inclusions have been observed which demonstrate intense immunoreactivity for both phosphorylated and non-phosphorylated neurofilament epitopes (Shaw et al. 1997b; Fig. 2C).

The primary pathogenetic processes underlying ALS are likely to be multifactorial, and the precise molecular mechanisms underlying selective cell death in the disease are at present unknown. Current understanding of the neurodegenerative process in ALS suggests that two major mechanisms contribute to motor-neurone injury: (a) glutamatergic toxicity and (b) oxidative stress/free radical damage (Brown 1995; Coyle and Puttfarcken 1993; Ince et al. 1997).

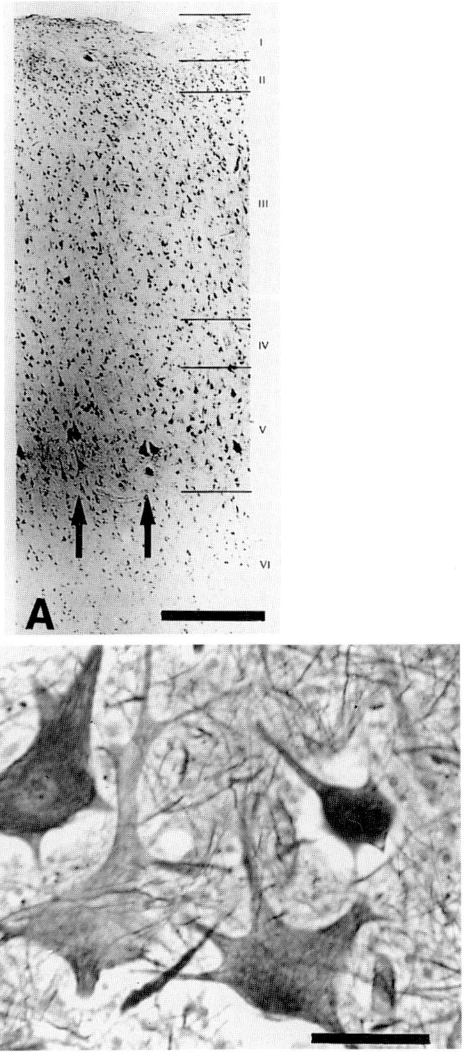

Fig. 1A,B. Normal human motor neurones. **A** *Arrows*, upper motor neurones or giant cells of Betz in the fifth layer of the normal human motor cortex. *Bar*, 1 mm. **B** Lower motor neurones in the ventral horn of the normal human lumbar spinal cord. *Bar*, 50 μm

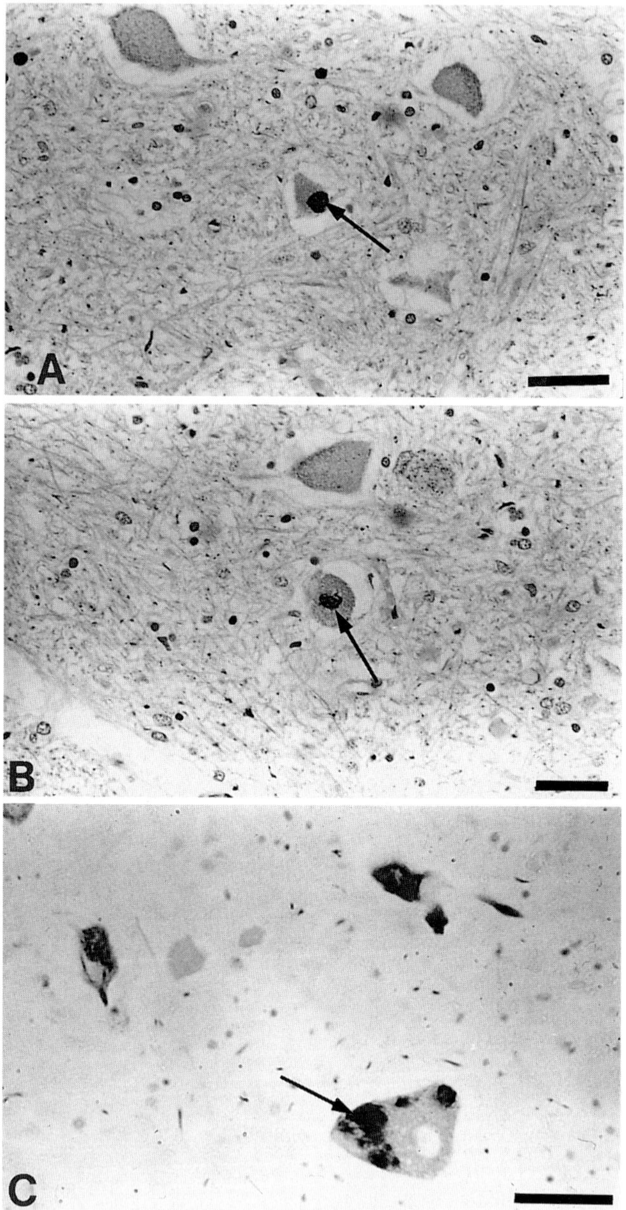

Fig. 2A–C. Legend see p. 69

This contribution highlights recent scientific developments in ALS under the following headings:

- The evidence for a disturbance of the glutamate neurotransmitter system in ALS
- Current knowledge about the molecular profile of glutamate receptors on human motor neurones
- The cellular effects of Cu/Zn superoxide dismutase (SOD_1) mutations
- The evidence for oxidative stress as a pathogenetic mechanism in sporadic ALS
- Factors underlying selective vulnerability of motor neurones to injury in ALS
- Links between glutamate and oxidative stress in neuronal injury

4.2 Evidence for Disturbance of Glutamate Neurotransmission in ALS

4.2.1 Normal Glutamate Neurotransmission

During normal glutamatergic neurotransmission, there is release of glutamate from pre-synaptic neuronal terminals and activation of post-synaptic receptors. The excitatory signal is terminated by active removal of glutamate from the synaptic cleft by several types of glutamate re-uptake transporter proteins which are located on peri-synaptic astrocytes and neurones. To date, four human glutamate transporter genes have been cloned: excitatory amino-acid transporter (EAAT): EAAT1 which is a glial transporter localised preferentially to the cerebellar cortex, EAAT2 which is a widely expressed glial-specific transporter,

◀ **Fig. 2A–C.** Inclusion bodies within motor neurones in cases with amyotrophic lateral sclerosis. **A** Hyaline inclusion (*arrow*) labelled with an antibody to ubiquitin. *Bar*, 100 µm. **B** Skein-like ubiquitinated inclusion (arrow). *Bar*, 100µm. **C** Hyaline conglomerate inclusions (*arrows*) in a case of familial ALS with an I113 T mutation in the SOD1 gene. These inclusions stain with antibodies to both phosphorylated and non-phosphorylated neurofilaments, as well as ubiquitin. *Bar*, 50 µm

EAAT3 which is localised to neurones, and EAAT4 which is expressed mainly in the cerebellum (Shashidharan and Plaitakis 1993; Shashidharan et al. 1994a,b; Arriza et al. 1994; Fairman et al. 1995). Synaptic glutamate which has been transported into peri-synaptic glial cells is converted to glutamine by the enzyme glutamine synthetase. Glutamine is then shuttled back to the neuronal terminal where it is reconverted to neurotransmitter glutamate by the action of glutaminase (Laake et al. 1995).

Of relevance to ALS is the probability that glutamate is an important excitatory neurotransmitter in several pathways in the human motor system. Glutamatergic inputs to motor neurones arise from the descending corticospinal pathways (Young et al. 1983; Young and Penney 1992), collaterals of the Aα fibres innervating muscle spindles and Golgi tendon organs (Burke 1990; Molander et al. 1989) and from spinal cord excitatory interneurones (O'Brien and Fischbach 1986).

The potential contribution of glutamate to motor-neurone injury in ALS has been the subject of several recent reviews (Shaw 1994; Zeman et al. 1994; Rothstein 1995; Shaw and Ince 1997). The key points can be summarised as follows.

4.2.1.1 Experimental Studies

The potential role of glutamate in causing motor-neurone toxicity has been shown in several experimental systems. For example, the CSF of ALS patients has been shown to be toxic to rat cortical neurones in culture, whereas the CSF of controls is not (Couratier et al. 1993). Intrathecal injection of the glutamate receptor agonist, kainic acid, in mice preferentially injures spinal motor neurones and induces the formation of abnormally phosphorylated neurofilaments, a cytoskeletal abnormality that has been observed in ALS (Hugon and Vallat 1990). Thirdly, it has been shown using a tissue culture model in which rat spinal cord explants are maintained under conditions of chronic glutamate uptake inhibition, that motor-neurone toxicity is produced with a subacute time course (Rothstein et al. 1993). Non-*N*-methyl—aspartate (NMDA) and non-α-amino-3-hydroxy 5-methyl-4-isoxazole propionic acid (AMPA) receptor antagonists show selective efficacy in preventing damage to motor neurones in these experimental models of glutamate-mediated injury.

4.2.1.2 Glutamate Levels in the CSF and CNS Tissue and Glutamate Transporter Function in ALS

The work of several groups has suggested that there may be an underlying defect in the metabolism, storage or transport of glutamate in ALS. Several studies have shown a decrease in the level of glutamate in various regions of post-mortem CNS tissue in ALS (Plaitakis et al. 1988; Perry et al. 1987; Tsai et al. 1991). The level of glutamate in the CSF, which reflects the extracellular level, has been reported to be significantly increased in ALS patients (Rothstein et al. 1990; Shaw et al. 1995a) although not all studies have confirmed this finding (Perry et al. 1990). The elevation of CSF glutamate may be present in only a subset of ALS patients (Shaw et al. 1995a,b). Whether this represents a subgroup of ALS patients with a distinct pathophysiology, and whether it has implications for therapy with anti-glutamate agents requires further investigation. It is recognised that there can be technical difficulties in accurately measuring glutamate levels in biological samples (Ferrarese et al. 1993), and the findings in relation to CSF glutamate require confirmation in future studies using larger groups of patients and employing stringent precautions to prevent non-specific changes in the levels of glutamate.

There is a body of evidence indicating that the expression and function of the glial glutamate reuptake transporter system is abnormal in ALS. Such alterations could contribute to excitotoxic neuronal injury by reducing the efficiency of synaptic clearance of glutamate and allowing an increase in the level of extracellular glutamate. Human motor neurones vulnerable to pathology in ALS are surrounded by a dense network of processes immunoreactive for the glial glutamate transporter EAAT2, whereas spared motor-neurone groups, for example, oculomotor neurones, have much less conspicuous perisomatic expression of this protein (Milton et al. 1997; Fig. 3A,B). Rothstein et al. (1992) showed in synaptosomal preparations taken from post-mortem tissue a specific defect in the Na^+-dependent glutamate transporter system in the spinal cord and motor cortex in ALS. Autoradiographic studies in ALS spinal cord showed a reduction in [^{3}H]D-aspartate binding to the glutamate reuptake transport system (Shaw et al. 1994a). Immunoblotting studies using anti-peptide polyclonal antibodies to assess the level of glutamate transporter proteins in homogenates of CNS tissue have shown a substantial reduction in the glial transporter EAAT2 in the spinal cord and

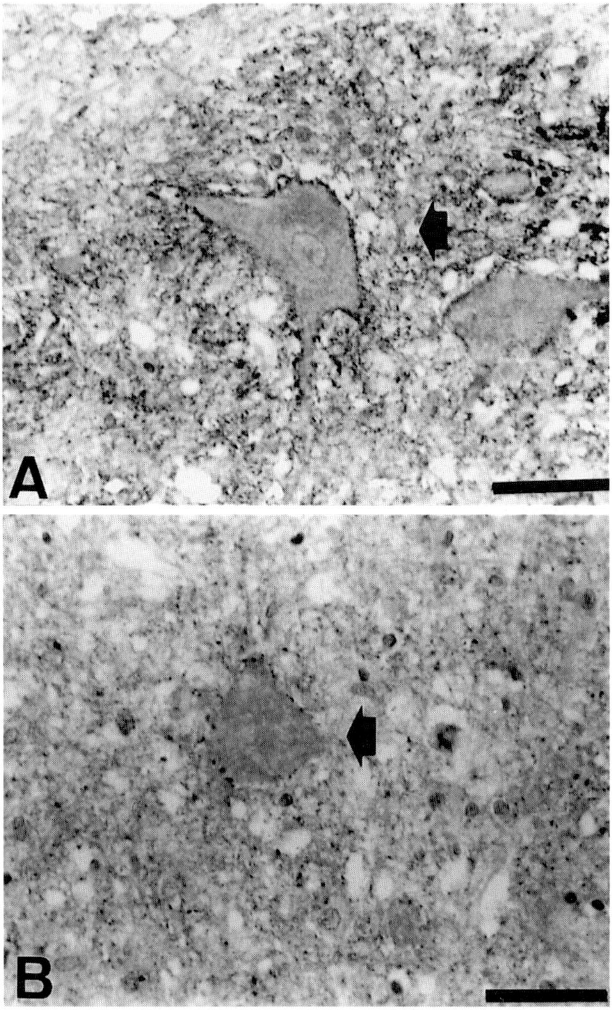

Fig. 3A,B. The expression of the glutamate transporter EAAT2 in the vicinity of normal human motor neurones. **A** The ventral horn of the lumbar spinal cord. There is strong expression of EAAT2 in the ventral horn neuropil and a dense rim of EAAT2 immunoreactivity in relation to the cell bodies of spinal motor neurones. *Bar*, 50 μm. **B** In contrast the expression of EAAT2 in the vicinity of oculomotor neurones, which are selectively spared in ALS, is much less strong. *Bar*, 50 μm motor cortex of ALS patients (Rothstein et al. 1995). In 25% of cases there was a very profound loss of EAAT2 expression.

A recent study using a monoclonal antibody to EAAT2 investigated the in situ expression of the EAAT2 protein in ALS and control cases using quantitative immunohistochemistry (Fray et al. 1997). The reduced expression of EAAT2 in ALS spinal cord was confirmed, but no reduction was observed in the motor cortex of ALS cases. The differences between these two studies may be due to differences in the antigenic specificity of the antibodies employed, differences in tissue retrieval and storage, as well as the potential instability of EAAT2 in tissue extracts. However, both studies provide evidence for an alteration in EAAT2 expression in ALS. No change in the expression of mRNA for EAAT2 has been observed in the motor cortex of ALS cases, suggesting that the loss of protein expression is likely to reflect a translational or post-translational effect rather than an alteration in gene structure, regulation or transcription (Bristol and Rothstein 1996). Further investigation is required to establish the mechanisms underlying the altered expression of EAAT2 in ALS. It is of interest that EAAT2 (murine equivalent GLT-1) is a protein known to be particularly vulnerable to damage by free radical species (Volterra et al. 1994).

The levels of GLT-1 (equivalent to human EAAT2) have been shown by immunoblotting to be decreased by 50% in spinal cord extracts from transgenic mice bearing the G85R human Cu/Zn superoxide dismutase (SOD_1) mutation (Bruijn et al. 1997). Even if the alteration of EAAT2 expression in ALS is a secondary process, it may still contribute to the cascade causing motor-neurone injury. In addition, therapy aimed at ameliorating the effects of altered glutamate transporter expression and function may contribute to the preservation and neuroprotection of motor neurones.

4.2.1.3 Alteration of Glutamate Receptor Expression and Function in ALS/Motor Neuron Disease

Autoradiographic studies have shown an increased density of binding sites for NMDA and non-NMDA receptor ligands in ALS, particularly in the intermediate grey matter of the spinal cord and deep layers of the motor cortex (Shaw et al. 1994b,c). This may reflect an increased excitatory drive to surviving motor neurones. Positron emission tomography scanning studies have shown abnormally widespread activation of several contralateral cortical regions, following freely selected upper limb movements (Kew et al.1993). This implies an inappropriate activation of neurones outside the somatotopic representation of the upper limb, suggesting an imbalance between cortical excitatory and inhibitory neurotransmission in individuals with ALS. Transcranial magnetic stimulation of the motor cortex in ALS has shown abnormalities in a proportion of patients reflecting the presence of hyperexcitability of motor neurones (Eisen et al. 1993a; Mills 1995).

4.2.1.4 Anti-glutamate Therapy in ALS and in Animal Models of the Disease

There is evidence for a modest improvement in survival with the anti-glutamate agent riluzole, both in human patients with ALS (Bensimon et al. 1994; Lacomblez et al. 1996) and in a transgenic mouse model of familial SOD_1 related ALS (Gurney et al. 1996). This therapeutic evidence suggests that abnormal glutamatergic mechanisms contribute to motor-neurone injury. Preliminary trials of other anti-glutamate agents including lamotrigine and gabapentin have not shown any evidence of therapeutic benefit to date, although the doses employed may have been suboptimal (Eisen et al. 1993b; Miller et al. 1996).

4.3 The Expression of Glutamate Receptor Subtypes by Human Motor Neurones

Motor neurones have a high density of glutamate receptors (Shaw et al. 1991; Williams et al. 1996) and in culture are susceptible to toxic effects following activation of either NMDA or non-NMDA receptors. Autoradiographic studies with specific radioligands for NMDA, AMPA and kainate receptor subtypes in the human spinal cord, brainstem and motor cortex have confirmed that these receptor subtypes are present in the motor-neurone areas (Shaw et al. 1991; Shaw and Ince 1994; Chinnery et al. 1993). Labelling of NMDA receptors with [^{3}H]MK-801 revealed a high concentration of these receptors as hot spots in close proximity to spinal motor-neurone cell bodies (Fig. 4A). With ligands for non-NMDA receptors there is a more uniform labelling of the ventral horn grey matter (Fig. 4B). Motor-neurone groups that tend to be spared in ALS (e.g. the oculomotor nucleus in the brainstem) show a lower

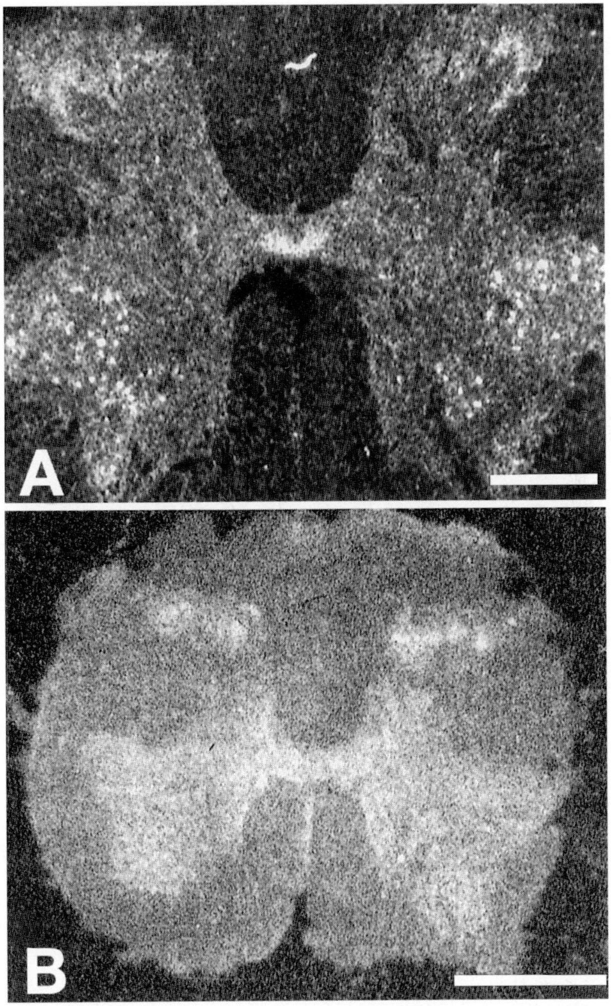

Fig. 4A,B. The expression of glutamate receptor subtypes in normal human spinal cord shown by autoradiography. **A** NMDA receptors in the lumbar spinal cord visualised using [^{3}H]MK-801. Focal areas of high binding are seen corresponding to the somata of lower motor neurones in the ventral horns. *Bar*, 1 mm. **B** AMPA receptors in the normal human lumbar spinal cord visualised using [^{3}H]CNQX. *Bar*, 3 mm

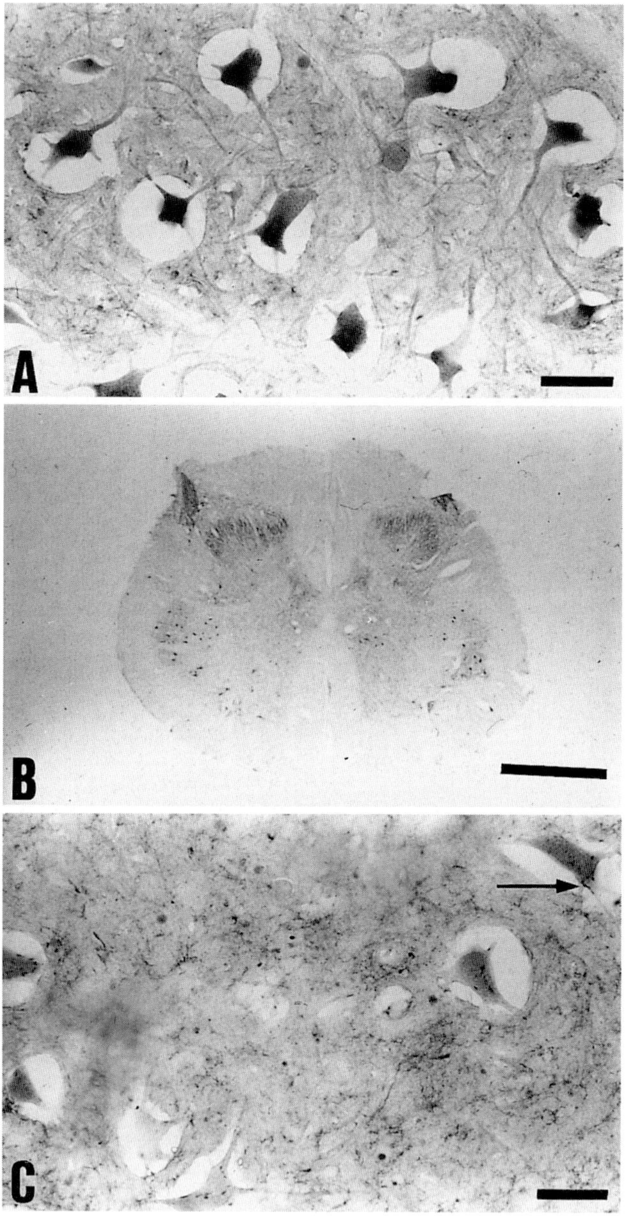

Fig. 5A–C. Legend see p. 77

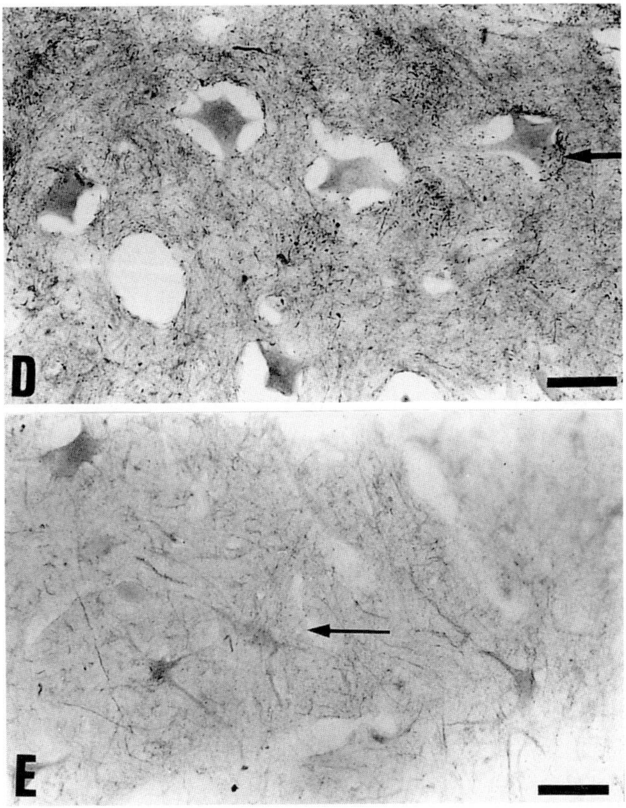

◀ **Fig. 5A–E.** The expression of glutamate receptor subunits by human spinal motor neurones assessed by immunohistochemistry using subunit specific antibodies. **A** NMDAR1 immunoreactivity. Motor neurones have very high expression of the universal NMDA receptor subunit NMDAR1. Unlike many other groups of neurones, the dendritic staining is much lighter than that of the cell body. *Bar*, 100 µm. **B** The expression of GluR2/3 AMPA receptor subunits in the normal human lumbar spinal cord. Strong immunoreactivity is observed in relation to the motor neurones and the substantia gelatinosa. *Bar*, 2 mm.
C Immunoreactivity to the kainate receptor subunit KA2. Motor neurones have a low-level diffuse immunoreactivity. Also observed are areas of dense immunoreactivity in a granular pattern, or sometimes visible as linear processes which have an appearance suggestive of pre-synaptic processes (*arrow*). *Bar*, 100 µm. **D** Kainate receptor subunits detected with an antibody to GluR6/7. There is moderate diffuse immunoreactivity of the motor-neurone cell bodies with some darkly staining beaded profiles on the cell surface (*arrow*). The neuropil contains beaded spots of staining which appear to be sited along neuronal processes. *Bar*, 100 µm. **E** The GluR2 AMPA receptor subunit. The majority of human motor neurones show low/absent immunoreactivity (*arrow*). *Bar*, 100 µm

density of NMDA receptor binding sites and a higher density of AMPA binding sites than motor-neurone groups vulnerable to the disease. These findings suggest differences in normal glutamatergic neurotransmission in spared and vulnerable motor-neurone groups.

More recent immunohistochemical studies using antibodies to specific glutamate receptor subunits have extended these findings. Human spinal motor neurones have a high expression of the universal NMDA receptor subunit $NMDAR_1$, and interestingly the dendritic staining is much lighter than the staining of the cell body, unlike many other groups of neurones (Fig. 5A). With AMPA receptor antibodies, motor neurones have a high expression of $GluR_{2/3}$ (Fig. 5B), an intermediate expression of $GluR_4$ and low expression of $GluR_1$. Spinal motor neurones have a diffuse low-level cytoplasmic immunoreactivity to the KA2 kainate receptor subunit. Also observed are irregular areas of dense immunoreactivity in a granular pattern, sometimes visible as linear processes which may well represent pre-synaptic processes (Fig. 5C). With an antibody to $GluR_{6/7}$ there is again diffuse cytoplasmic staining of motor-neurone somata, with dark beaded profiles on the cell surface. The neuropil contains beaded spots of $GluR_{6/7}$ staining which appear to be sited along neuronal processes (Fig. 5D).

Some recent data have emerged with respect to the expression of molecular subtypes of AMPA receptor subunits by human motor neurones. AMPA receptors are one of the major classes of ionotropic glutamate receptors and are responsible for much of the fast excitatory neurotransmission in the mammalian CNS. AMPA receptors are composed of heteromeric complexes of four protein subunits, shown as $GluR_1$–$GluR_4$. The $GluR_2$ subunit is of particular functional importance because when incorporated into the receptor complex in its dominant edited form, it renders the AMPA receptor complex impermeable to calcium (Burnashev et al. 1992). Most AMPA receptors in the human CNS include the $GluR_2$ subunit and are impermeable to calcium (Bettler and Mulle 1995; Day et al. 1995). Only a few groups of cells in the mammalian CNS appear to express calcium-permeable AMPA receptors. These include Bergmann glia in the cerebellum and a subpopulation of hippocampal neurones (Bettler and Mulle 1995; Westbrook 1994). Because of the importance of calcium in mediating the downstream toxic effects of glutamate, neuronal groups lacking $GluR_2$ expression would be expected to be more vulnerable to excitotoxicity.

Indeed, it has been shown that neuronal subpopulations expressing atypical AMPA receptors that lack $GluR_2$ exhibit heightened vulnerability to non-NMDA agonist toxicity (Brorson et al. 1995).

Using riboprobe in situ hybridisation to study the expression of mRNA for AMPA receptor subunits it has been shown that upper and lower motor neurones in the human CNS have low to absent signal for $GluR_2$ mRNA, whereas there is discernible expression of mRNA for $GluR_1$, $GluR_2$ and $GluR_3$ (Williams et al. 1997). This finding has been confirmed at protein level using an antibody specific for the $GluR_2$ AMPA receptor subunit recently developed in the laboratory of R. Wenthold (Fig. 5E).

The physiological significance of the expression of atypical Ca^{2+}-permeable AMPA receptors by certain groups of neurones is at present unknown. However, it is possible that the expression of Ca^{2+}-permeable AMPA receptors lacking $GluR_2$ by human motor neurones could render them selectively vulnerable to glutamate mediated injury by increasing calcium influx during receptor activation.

4.4 The Cellular Effects of Cu/Zn Superoxide Dismutase (SOD_1) Mutations

Oxidative stress is one of the major potential causes of age-related deterioration in neuronal function and cumulative damage by free radicals may contribute to the delayed onset and progressive nature of human neurodegenerative diseases. ALS is familial with an autosomal dominant pattern of inheritance in 5%–10% of cases. Genetically determined abnormalities in the cellular free radical defence system caused by mutations in the Cu/Zn superoxide dismutase (SOD_1) enzyme underlie 20% of all cases of familial ALS (Rosen et al. 1993; Radunovic and Leigh 1996). The primary role of SOD_1 is to catalyse the dismutation of superoxide radicals to hydrogen peroxide (H_2O_2) which is then eliminated by glutathione peroxidase or catalase. Subsidiary activities include peroxidase activity resulting in the generation of hydroxyl radicals from H_2O_2 (Wiedau-Pazos et al. 1996), the production of nitronium species from peroxynitrite (Beckman et al. 1993) and the protection of the enzyme calcineurin from inactivation (Wang et al. 1996).

To date approximately 47 different mutations have been described involving exons 1, 2, 4 and 5 (Siddique et al. 1996) and non-coding sequences (Shaw et al. 1997a,b), but sparing exon 3, which encodes the active site of the enzyme. The sites of the SOD_1 protein affected by the described mutations tend to affect the dimer stability or β barrel folding of the enzyme (Brown 1995). SOD_1 activity assayed in red blood cells in transformed lymphoblastoid cells and in brain tissue from patients with SOD_1 mutations is 30%–70% of the level found in controls. This may result from reduced stability and half-life of the mutant protein (Borchelt et al. 1995).

The mechanisms of selective motor-neurone injury in the presence of SOD_1 mutations remain uncertain. It has been established that normal motor neurones have a higher expression of SOD_1, in both the somatic and axonal compartments, than do other groups of neurones in the human CNS (Pardo et al. 1995; Shaw et al. 1997a,b). It has not been established with certainty whether a loss of function of SOD_1 or a toxic gain of function of mutant SOD_1 is responsible for motor-neurone injury, nor is it established that all SOD_1 mutations have the same physiological effects. The loss of function hypothesis postulates that degeneration of motor neurones occurs by the direct effect of superoxide radicals not properly scavenged by mutant SOD_1. This hypothesis is favoured by the following lines of evidence: (a) There are a large number of mutations in SOD_1 which cause a decrease in cytosolic enzyme activity but are not predicted to share similar structural changes (Siddique et al. 1996). (b) Mutant SOD_1 in Drosophila causes a dominant negative effect on the normal SOD_1 protein (Phillips et al. 1995). (c) In organotypic spinal cord cultures and cultures of PC12 cells, reduction of SOD_1 activity by chronic application of anti-sense oligonucleotides triggers cell death by apoptosis (Troy and Shelanski 1994; Rothstein et al. 1994).

However, there is a body of scientific evidence including observations from transgenic animals (discussed below) which favours the possibility that the mutant SOD_1 molecule exerts a toxic gain of function which is responsible for motor-neurone injury. Several hypotheses have been put forward to explain this toxic gain of function including: (a) *Abnormal metal binding* of the mutant enzyme which could cause Cu^{2+} or Zn^{2+} toxicity to motor neurones. (b) *Toxic protein aggregation;* it has been shown in primary neuronal cultures microinjected with mutant SOD_1 that these cells accumulate aggregates of SOD_1 (Durham et al. 1997). (c) *Altered substrate affinity;* many of the missense mutations found in SOD_1 have the potential to alter the active site of the SOD_1 enzyme. Such changes could enable mutant SOD_1 to react with additional substrates such as hydrogen peroxide and peroxynitrite (Brown 1995; Beckman et al. 1993). These aberrant reactions could result in the production of highly reactive and damaging hydroxyl radicals and nitronium ions which can damage proteins by nitration of tyrosine residues. Neurofilament proteins and neurotrophic factor (tyrosine kinase) receptors, both critical for the normal function of motor neurones, are proteins particularly vulnerable to nitrotyrosine damage (Beckman et al. 1993).

In vitro studies from the laboratory of Bredesen et al. have shown that at least two of the SOD_1 mutations can increase the rate of hydroxyl radical formation, probably due to increased availability of the active copper site to H_2O_2 (Wiedau-Pazos et al. 1996). This group also showed that when expressed in a neuronal cell line, the mutant SOD_1 proteins promoted apoptosis induced by serum or growth factor withdrawal, whereas over-expression of wild-type SOD_1 inhibited apoptosis (Rabizadeh et al. 1995).

4.4.1 SOD_1 Transgenic Mice

Three transgenic SOD_1 mouse strains, G93A, G85R and G37R have been shown to develop an ALS-like phenotype (Gurney et al. 1994; Ripps et al. 1995; Wong et al. 1995), whereas mice with wild-type human SOD_1 expressed at similar levels do not develop similar features. The mutant SOD_1 animals develop progressive weakness and wasting affecting limb and bulbar muscles. The disease process usually starts between 3 and 8 months of age and is usually fatal within a few weeks. The spinal and brainstem lower motor neurones are primarily affected. The different mutations produce somewhat different pathological effects. The G93A and the G37R transgenic mice tend to develop early changes of vacuolation in dendrites and axons of motor neurones, with reactive gliosis, followed by vacuolation of the somata and cell death. There is prominent swelling and vacuolation of mitochondria and early fragmentation of the Golgi apparatus. The G93A mutant develops neurofilament abnormalities akin to human ALS. The G85R mutant develops early pathological changes within astrocytes that stain with antibodies to SOD_1 and ubiquitin (Bruijn et al. 1997). These changes are accompanied by reduced expression of the glial glutamate transporter GLT-1. It has recently been shown that G93A mice expressing a low copy number of the transgene develop pathological changes most closely resembling those of human motor neuron disease (Dal Canto and Gurney 1995; Tu et al. 1996). In these mice there is motor-neurone degeneration associated with neuronal and axonal filamentous inclusions, immunoreactive for neurofilaments and ubiquitin, together with astrocytic inclusions.

A SOD_1 knockout mouse has been produced which does not develop an ALS phenotype. However, motor neurones in this mouse model are more vulnerable to cell death following injury such as axotomy (Reaume et al. 1996).

Data are beginning to emerge on the effect of potential neuroprotective strategies in SOD_1 transgenic mice. In the Gurney G37R mouse the anti-oxidant vitamin E delays the onset of clinical disease but does not extend survival, whereas the anti-glutamate agent riluzole extends survival by 11% but does not delay the onset of disease (Gurney et al. 1996). These findings suggest that different mechanisms are important at different stages of the disease, i.e. that toxic effects of glutamate may occur later than oxidative stress. Friedlander et al. (1997) recently crossed G93R SOD_1 transgenic mouse with a transgenic mouse expressing a dominant negative inhibitor of the interleukin-1β-converting enzyme (ICE). They showed that inhibition of this enzyme is able to slow symptomatic progression of the murine motor-neurone disease and delay mortality.

4.5 Oxidative Stress in Sporadic ALS

There is close clinical and pathological similarity between familial and sporadic ALS, which suggests that they may share common pathophysiological mechanisms. Several lines of evidence have emerged which suggest that oxidative stress contributes to motor-neurone injury in sporadic ALS.

4.5.1 Biochemical Indices of Oxidative Damage

Protein carbonyls, which are formed by oxidative modification of certain amino acid residues in protein are elevated in the spinal cord (Shaw et al. 1995a,b) and frontal cortex of ALS patients (Bowling et al. 1993).

4.5.2 Biochemical Indices Which May Indicate a Compensatory Response to Oxidative Stress

The content of selenium and the activity of the selenium-containing free radical scavenging enzyme glutathione peroxidase are higher in the spinal cord of ALS cases than in controls (Ince et al. 1994). SOD_1 mRNA has been shown to be increased in individual motor neurones from sporadic ALS cases (Bergeron et al. 1994). Immunohistochemical studies in human spinal cord have shown increased glial immunoreactivity for SOD_1 Mn SOD and catalase in ALS in the vicinity of the corticospinal tracts and/or in the neuropil of the ventral horn (Shaw et al. 1997a,b). Increased expression has been demonstrated in ALS spinal cord of metallothionein (Sillevis-Smith et al. 1994). Metallothioneins are metal binding proteins involved in the detoxification and storage of metals and have free radical scavenging capabilities.

4.5.3 Candidate Targets for Free Radical Damage in the Human Motor System

4.5.3.1 Glial Proteins Important in the Regulation of Glutamate Neurotransmission

Glutamate synthetase (GS) is an astrocytic enzyme essential for the recycling of neurotransmitter glutamate. It converts glutamate to glutamine which is then exported to neurones for recycling to neurotransmitter glutamate (Laake 1995). A reduction in GS activity results in an increase in extracellular glutamate and a decrease in neuronal glutamate (Schouseboe et al. 1993), neurochemical changes similar to those reported in ALS. Glial glutamate transporter proteins EAAT1 and EAAT2 are essential for the removal of glutamate from the synaptic cleft and the termination of its excitatory signal. In experimental systems both GS and EAAT2 have been shown to be especially sensitive to oxidative damage (Volterra et al. 1994; Schor 1988).

4.5.3.2 Neurofilament Proteins

Neurofilament proteins are of interest for several reasons. Firstly, the hyaline ubiquitinated inclusion bodies present in motor neurones in ALS may contain neurofilament epitopes as shown by double immunofluorescent labelling and confocal microscopy (Thatcher et al. 1996). Secondly, certain familial cases with particular SOD_1 mutations have dramatic inclusion bodies which are strongly reactive for both phosphorylated and non-phosphorylated neurofilament proteins (Shaw et al. 1997b). Finally, in neuronal cell culture models of oxidative stress neurofilament proteins appear to show differential vulnerability to free radical damage (Cookson et al. 1996).

4.5.3.3 Mitochondrial Proteins

Mitochondria are the major intracellular sites of free radical generation via the reactions of the electron transport chain. Mitochondria have been shown to be particularly susceptible to free radical damage at both protein and DNA levels (Linnane et al. 1989; Richter et al. 1988), and free radicals are known to inhibit the activities of specific mitochondrial enzymes (Zhang 1990). The most sensitive mitochondrial proteins appear to be NADH dehydrogenase, ATPase and succinate dehydrogenase. There is evidence that mitochondrial dysfunction has a role in ALS. Firstly samples of frontal cortex from cases of familial ALS showed alteration in the activity of the complex I component of the electron transport chain (Bowling et al. 1993). Secondly, vacuolar distortion of mitochondria has been observed in both the G93A and G37R SOD_1 transgenic mice at an early stage in the pathological process, prior to the appearance of clinical signs (Gurney et al. 1994; Chiu et al. 1995).

4.6 Factors Underlying Selective Vulnerability of Motor Neurones to Injury in ALS

Motor neurones are amongst the largest cells in the human CNS. They have large cell bodies and may have to maintain an axonal system up to 1 m in length. To achieve this they have a high metabolic rate and a robust cytoskeleton with a high content of neurofilament proteins. The high expression of SOD_1 in normal human motor neurones may reflect the high metabolic activity of this cell group. It is possible that this high content of SOD_1 predisposes motor neurones to selective injury when the SOD_1 protein is mutated (Pardo et al. 1995).

Several molecular features of human motor neurones are beginning to be uncovered, which may explain why these cells might be vulnerable to glutamate mediated injury. Firstly, as discussed above, human motor neurones have a low/absent expression of the AMPA receptor subunit $GluR_2$ mRNA and protein (Williams et al. 1997). This implies that motor-neurone AMPA receptors are be predominantly Ca^{2+} permeable. Secondly, motor neurones which are vulnerable to the pathological effects in ALS, do not express the calcium-binding proteins parvalbumin and calbindin D28 K (Ince et al. 1993; Alexianu et al. 1994). This is in contrast to groups of motor neurones such as the oculomotor neurones which tend to be spared in ALS, and which do express parvalbumin. One of the functions of these proteins is to buffer intracellular calcium, and neurones which lack their expression may be more vulnerable to toxic effects of calcium following glutamate receptor activation. Thus, two molecular features of motor neurones, the lack of $GluR_2$ AMPA receptor subunit and the lack of calcium buffering proteins, may render them excessively vulnerable to calcium-mediated toxic effects following activation of their cell surface glutamate receptors.

4.7 Links Between Glutamate Neurotransmission and Oxidative Stress

The two potential mechanisms of motor-neurone injury discussed in this review – glutamatergic mechanisms and oxidative stress – have important links with each other and may act as co-conspirators in the cell death process (Coyle and Puttfarcken 1993). Activation of glutamate receptors is one of the main mechanisms for intracellular generation of free radicals via calcium-dependent activation of the arachidonic acid cascade, nitric oxide synthase and calpain/xanthine oxidase (Lees 1993; Pellegrini-Giampietro 1994). Reactive oxygen species can cause damage to intracellular proteins, lipids and DNA, with cumulative impairment of the function of essential macromolecules and organelles. Some proteins which appear particularly sensitive to free radical damage including EAAT2 (GLT-1) and GS, as discussed above, are very important in the regulation of glutamate neurotransmission.

Glutamate can also indirectly affect the ability of neurones to withstand oxidative stress by causing depletion of intracellular glutathione (Murphy et al. 1989). The cystine carrier transports cystine into the cell and counter-transports glutamate out of the cell, with the glutamate gradient providing the driving force for the exchange. Cystine and glutamate compete for binding to the carrier and an increase in extracellular glutamate result in decreased intracellular cystine transport. Cystine is an important precursor of glutathione which is a vital component of intracellular defence against oxidative stress (Meister and Anderson 1983).

Mitochondria are the most important source for the formation of oxygen free radicals within neurones. It has been demonstrated that mitochondrial production of free radicals increases in cultured neurones exposed to glutamate (Dugan et al. 1994). Age-related changes in mitochondrial function have been suggested as a major factor contributing to neurodegenerative disease of late onset. Accumulating mutations in the mitochondrial genome appear important in the genesis of age-related deterioration in mitochondrial function (Richter 1995). Mitochondrial DNA is particularly susceptible to oxidative damage due to factors such as proximity to the major intracellular source of free radical generation, minimal DNA repair mechanisms, the absence of protective histones, and the low ratio of non-coding to coding sequences compared to nuclear DNA.

Normal mitochondrial function is essential for maintenance of tonic Mg^{2+} blockade of NMDA glutamate receptor channels and for intracellular calcium buffering (Linnane et al. 1989; White and Reynolds 1995). These properties may explain why impairment of mitochondrial function lowers the threshold for glutamate-induced neuronal injury (Beal 1993). Thus, once the neuronal cell falls below a certain critical threshold in energy production, it becomes susceptible to further injury by secondary excitotoxicity.

4.8 Conclusion

In conclusion, progress is beginning to be made in unravelling the molecular mechanisms which contribute to motor-neurone injury in ALS. The evidence to date suggests that the cell death process in ALS reflects a complex interaction between genetic factors, toxic activation of glutamate receptors, oxidative stress and damage to critical target molecules and organelles. The relative importance of these pathways may vary in subgroups of ALS patients, and unravelling the subgroups will prove a very important task for the future.

Acknowledgements. P.J.S. is supported by the Wellcome Trust as a Senior Research Fellow in Clinical Science.

References

Alexianu ME, Ho BK, Mohamed AH, La Bella V, Smith RG, Appel S (1994) The role of calcium-binding proteins in selective motoneuron vulnerability in amyotrophic lateral sclerosis. Ann Neurol 36:846–858

Arriza JL, Fairman WA, Wadiche JI, Murdoch GH, Kavanaugh MP, Amara SG (1994) Functional comparisons of three glutamate transporter subtypes cloned from human motor cortex. J Neurosci 14(9):5559–5569

Beal MF, Hyman BT, Koroshetz W (1993) Do defects in mitochondrial energy metabolism underlie the pathology of neurodegenerative diseases. Trends Neurol Sci 16:125–131

Beckman JS, Carson M, Smith CD, Koppenol WH (1993) ALS, SOD and peroxynitrite. Nature 364:584

Bensimon G, Lacomblez L, Meininger V et al (1994) A controlled trial of riluzole in amyotrophic lateral sclerosis. New Engl J Med 330:585–591

Bergeron C, Muntasser S, Somerville MJ, Weyer L, Percy ME (1994) Copper/zinc superoxide dismutase mRNA levels are increased in sporadic amyotrophic lateral sclerosis motorneurons. Brain Res 659(1–2):272–276

Bettler B, Mulle C (1995) Review: neurotransmitter receptors II AMPA and kainate receptors. Neuropharmacology 34:123–139

Borchelt DR, Guarnieri M, Wong PC et al (1995) Superoxide dismutase 1 subunits with mutations linked to familial amyotrophic lateral sclerosis do not affect wild-type subunit function. J Biol Chem 270(7):3234–3238

Bowling AC, Schulz JB, Brown RH, Beal MF (1993) Superoxide dismutase activity, oxidative damage and mitochondrial energy metabolism in familial and sporadic amyotrophic lateral sclerosis. J Neurochem 61:2322–2325

Bristol LA, Rothstein JD (1996) Glutamate transporter gene expression in amyotrophic lateral sclerosis motor cortex. Ann Neurol 39:676–679

Brorson JR, Manzolillo PA, Gibbons SJ, Miller RJ (1995) AMPA receptor desensitization predicts the selective vulnerability of cerebellar Purkinje cells to excitotoxicity. J Neurosci 15(6):4515–4524

Brown RHJ (1995) Amyotrophic lateral sclerosis: recent insights from genetics and transgenic mice. Cell 80:687–692

Bruijn LI, Becher MW, Lee MK et al (1997) ALS-linked SOD1 mutant G85R mediates damage to astrocytes and promotes rapidly progressive disease with SOD1 containing inclusions. Neuron 18:327–338

Burke RE (1990) Spinal cord: ventral horn. In: Shepherd GM (ed) The synaptic organisation of the brain. Oxford University Press, New York, pp 88–132

Burnashev N, Monyer H, Seeburg PH, Sakmann B (1992) Divalent ion permeability of AMPA receptor channels is dominated by the edited form of a single subunit. Neuron 8:189–198

Chari G, Shaw PJ, Sahgal A (1996) Non-verbal visual attention, but not recognition memory or learning processes are impaired in motor-neurone disease. Neuropsychologica 34:377–385

Chinnery RM, Shaw PJ, Ince PG, Johnson M (1993) Autoradiographic distribution of binding sites for the non-NMDA receptor antagonist [3H]CNQX in human motor cortex, brainstem and spinal cord. Brain Res 630:75–81

Chiu AY, Zhai P, Dal Canto MC et al (1995) Age-dependent penetrance of disease in a transgenic mouse model of familial amyotrophic lateral sclerosis. Mol Cell Neurosci 6:349–362

Cookson MR, Thatcher NM, Ince PG, Shaw PJ (1996) Selective loss of neurofilament proteins after exposure of differentiated IMR-32 neuroblastoma cells to oxidative stress. Brain Res 738:162–166

Couratier P, Hugon J, Sindou P, Vallat JM, Dumas M (1993) Cell culture evidence for neurone degeneration in amyotrophic lateral sclerosis being linked to AMPA/kainate receptors. Lancet 341:265–268

Coyle JT, Puttfarcken P (1993) Oxidative stress, glutamate and neurodegenerative disorders. Science 262:689–695

Dal Canto MC, Gurney ME (1995) Neuropathological changes in two lines of mice carrying a transgene for mutant human Cu, Zn SOD and in mice expressing wild type human SOD: a model of familial amyotrophic lateral sclerosis (FALS). Brain Res 676:24–40

Day NC, Williams TL, Ince PG, Kamboj RK, Lodge D, Shaw PJ (1995) Distribution of AMPA-selective glutamate receptor subunits in the human hippocampus and cerebellum. Mol Brain Res 31:17–32

Dugan LL, Sensi SL, Canzoniero LMT et al (1994) Imaging of mitochondrial oxygen radical production in cortical neurons exposed to NMDA. Soc Neurosci Abstr 19:21

Durham HD, Roy J, Dong L, Figlewicz DA (1997) Aggregation of mutant Cu/Zn superoxide dismutase proteins in a culture model of ALS. J Neuropathol Exp Neurol 56:523–530

Eisen A, Pant B, Stewart H (1993a) Cortical excitability in amyotrophic lateral sclerosis: a clue to pathogenesis. Can J Neurol Sci 20:11–16

Eisen A, Stewart H, Schulzer M, Cameron D (1993b) Anti-glutamate therapy in amyotrophic lateral sclerosis: a trial using lamotrigine. Can J Neurol Sci 20:11–16

Fairman WA, Vandenberg RJ, Arriza JL, Kavanaugh MP, Amara SG (1995) An excitatory amino-acid transporter with properties of a ligand-gated chloride channel. Nature 375:599–603

Ferrarese L, Pecora N, Frigo M, Appollonio I, Frattola L (1993) Assessment of reliability and biological significance of glutamate levels in cerebrospinal fluid. Ann Neurol 33:316–319

Fray AE, Banner SJ, Ince PG, Milton ID, Usher PA, Shaw PJ (1997) Expression of the glial glutamate transporter EAAT2 in motor-neurone disease: an immunocytochemical study. Eur J Neurosci (submitted)

Friedlander RM, Brown RH, Gagliardini V, Wang J, Yuan J (1997) Inhibition of ICE slows ALS in mice. Nature 388:31

Gurney ME, Pu H, Chiu AY (1994) Motor neuron degeneration in mice that express a human Cu/Zn superoxide dismutase mutation. Science 264:1772–1775

Gurney ME, Cutting FB, Zhai P et al (1996) Benefit of vitamin E, riluzole and gabapentin in a transgenic model of familial amyotrophic lateral sclerosis. Ann Neurol 39:147–158

Hugon J, Vallat JM (1990) Abnormal distribution of phosphorylated neurofilaments in neuronal degeneration induced by kainic acid. Neurosci Lett 119:45–48

Ince PG, Stout N, Shaw PJ et al (1993) Parvalbumin and calbindin D-28k in the human motor system and in motor neuron disease. Neuropathol Appl Neurobiol 19:291–299

Ince PG, Shaw PJ, Candy JM et al (1994) Iron, selenium and glutathione peroxidase activity are elevated in sporadic motor neuron disease. Neurosci Lett 183:87–90

Ince PG, Shaw, PJ, Slade JY, Jones C, Hudgson P (1996) Familial amyotrophic lateral sclerosis with a mutation in exon 4 of the Cu/Zn superoxide dismutase gene: pathological and immunocytochemical changes. Acta Neuropath 92:395–403

Ince PG, Eggett C, Shaw PJ (1997) Role of excitotoxicity in neurological disease. Reviews in Contemporary Pharmacology (in press)

Iwanaga K, Hayashi S, Oyake M et al (1997) Neuropathology of sporadic amyotrophic lateral sclerosis of long duration. J Neurol Sci 146:139–143

Kew JJM, Leigh PN, Playford ED et al (1993) Cortical function in amyotrophic lateral sclerosis. A positron emission tomography study. Brain 116:655–680

Laake JH, Slyngstad TA, Haug F-MS, Ottersen OP (1995) Glutamine from glial cells is essential for the maintenance of the nerve terminal pool of glutamate: immunogold evidence from hippocampal slice cultuRes J Neurochem 65:871–881

Lacomblez L, Bensimon G, Leigh PN et al (1996) Dose-ranging study of riluzole in amyotrophic lateral sclerosis. Lancet 347:1425–1432

Lees GJ (1993) Contributory mechanisms in the causation of neurodegenerative disorders. Neuroscience 54:287–322

Leigh PN, Anderton B, Dodson A, Gallo J-M, Swash M, Power D (1988) Ubiquitin deposits in anterior horn cells in motor neurone disease. Neurosci Lett 93:197–203

Linnane AW, Marzuki S, Ozawa T, Tanaka M (1989) Mitochondrial DNA mutations as an important contribution to ageing and degenerative diseases. Lancet :642–645

Lowe J, Lennox G, Jefferson D et al (1988) A filamentous inclusion within anterior horn cell neurones in motor neurone disease defined by immunocytochemical localisation with ubiquitin. Neurosci Lett 93:202–210

Meister A, Anderson ME (1983) Glutathione. Annu Rev Biochem 52:711–760

Miller RG, Moore D, Young LA (1996) Placebo-controlled trial of gabapentin in patients with amyotrophic lateral sclerosis. Neurology 47:1383–1388

Mills KR (1995) Motor neurone disease: studies of the corticospinal excitation of single motoneurons by magnetic brain stimulation. Brain 118:971–902

Milton ID, Banner SJ, Ince PG et al (1997) The immunohistochemical expression of the glial glutamate transporter EAAT2 in the human CNS. Mol Brain Res (in press)

Molander C, Xu Q, Rivero-Mellian C, Grant G (1989) Cytoarchitectonic organisation of the spinal cord in the rat. II. The cervical and thoracic cord. J Comp Neurol 1989:375–385

Murphy TH, Miyamoto M, Sastre A, Schnaar RL, Coyle JT (1989) Glutamate toxicity in a neuronal cell line involves inhibition of cystine transport leading to oxidative stress. Neuron 2:1547–1558

O'Brien RJ, Fischbach GD (1986) Characterisation of excitatory amino acid receptors expressed by chick motor neurons in vitro. J Neurosci 6:3290–3296

Pardo CA, Xu Z, Borchelt DR, Price DL, Sisodia SS, Cleveland DW (1995) Superoxide dismutase is an abundant component in cell bodies, dendrites and axons of motor neurons and in a subset of other neurons. Proc Natl Acad Sci USA 92(4):954–958

Pelligrini-Giampietro DE (1994) Free radicals and the pathogenesis of neuronal death: cooperative role of excitatory amino acids. In: Armstrong D (ed) Free radicals in diagnostic medicine. Plenum, New York, pp 59–71

Perry TL, Hansen S, Jones K (1987) Brain glutamate deficiency in amyotrophic lateral sclerosis. Neurology 37:1845–1848

Perry TL, Krieger C, Hansen S, Eisen A (1990) Amyotrophic lateral sclerosis: amino acid levels in plasma and cerebrospinal fluid. Ann Neurol 28:12–17

Phillips JP TJ, Getzoff ED et al (1995) Subunit-destabilising mutations in Drosophila copper/zinc superoxide dismutase: neuropathology and a model of dimer dysequilibrium. Proc Natl Acad Sci USA 92:8533–8534

Plaitakis A, Constantakakis E, Smith J (1988) The neuroexcitotoxic amino acids glutamate and aspartate are altered in the spinal cord and brain in amyotrophic lateral sclerosis. Ann Neurol 24:446–449

Rabizadeh S, Ralla EB, Borchelt DR et al (1995) Mutations associated with amyotrophic lateral sclerosis convert superoxide dismutase from an antiapoptotic gene to a proapoptotic gene: studies in yeast and neural cells. Proc Natl Acad Sci USA 92:3024–3028

Radunovic A, Leigh PN (1996) Cu/Zn superoxide dismutase gene mutations in amyotrophic lateral sclerosis: correlation between genotype and clinical features. J Neurol Neurosurg Psychiatry 61:565–572

Reaume A, Elliott JL, Hoffman EK et al (1996) Motor neurons in Cu/Zn superoxide dismutase-deficient mice develop normally but exhibit enhanced cell death after axonal injury. Nature Genet 13:43–47

Richter C (1995) Oxidative damage to mitochondrial DNA and its relationship to ageing. Int J Biochem Cell Biol 27:647–653

Richter C, Park JW, Ames BN (1988) Normal oxidative damage to mitochondrial and nuclear DNA is extensive. Proc Natl Acad Sci USA 85:6465–6467

Ripps ME, Huntley GW, Hof PR, Morrison JH, Gordon JW (1995) Transgenic mice expressing an altered murine superoxide dismutase gene provide an animal model of amyotrophic lateral sclerosis. Proc Natl Acad Sci USA 92(3):689–693

Rosen DR, Siddique T, Patterson D et al (1993) Mutations in Cu/Zn superoxide dismutase are associated with familial amyotrophic lateral sclerosis. Nature 362:59–62

Rothstein JD (1995) Excitotoxic mechanisms in the pathogenesis of amyotrophic lateral sclerosis. Adv Neurol 68:7–20

Rothstein JD, Tsai G, Kuncl RW (1990) Abnormal excitatory amino acid metabolism in amyotrophic lateral sclerosis. Ann Neurol 28:18–25

Rothstein JD, Martin LJ, Kuncl RW (1992) Decreased glutamate transport by the brain and spinal cord in amyotrophic lateral sclerosis. New Engl J Med 326:1464–1468

Rothstein JD, Jin L, Dykes-Hoberg M, Kuncl RW (1993) Chronic inhibition of glutamate uptake produces a model of slow neurotoxicity. Proc Natl Acad Sci 90:6591–6595

Rothstein JD, Bristol LA, Hosler B, Brown RH, Kuncl RW (1994) Chronic inhibition of superoxide dismutase produces apoptotic death of spinal neurons. Proc Natl Acad Sci USA 91:4155–4159

Rothstein JD, Van Kammen M, Levey AI, Martin LJ, Kuncl RW (1995) Selective loss of glial glutamate transporter GLT-1 in amyotrophic lateral sclerosis. Ann Neurol 38:73–84

Schor NF (1988) Inactivation of mammalian brain glutamine synthetase by oxygen radicals. Brain Res 456:17–21

Schousboe A, Westergaard N, Sonnewald U et al (1993) Glutamate and glutamine metabolism and compartmentation in astrocytes. Dev Neurosci 15:359–366

Shashidharan P, Plaitakis A (1993) Cloning and characterisation of a glutamate transporter cDNA from human cerebellum. Biochim Biophys Acta 1216:161–164

Shashidharan P, Huntley GW, Meyer T, Morrison JH, Plaitakis A (1994a) Neuron-specific human glutamate transporter: molecular cloning, characterization and expression in human brain. Brain Res 662:245–250

Shashidharan P, Wittenberg I, Plaitakis A (1994b) Molecular cloning of human brain glutamate/aspartate transporter II. Biochim Biophys Acta 1191:393–396

Shaw PJ (1994) Excitotoxicity and motor neurone disease: a review of the evidence. J Neurol Sci 124 [Suppl]:6–13

Shaw PJ, Ince PG, Johnson M, Perry EK, Candy JM (1991) The quantitative autoradiographic distribution of [3H]MK-801 binding sites in the normal human spinal cord. Brain Res 539:164–168

Shaw PJ, Ince PG (1994) A quantitative autoradiographic study of [3H]kainate binding sites in the normal spinal cord, brainstem and motor cortex. Brain Res 641:39–45

Shaw PJ, Chinnery RM, Ince PG (1994a) [3H]D-Aspartate binding sites in the normal human spinal cord and changes in motor neuron disease: a quantitative autoradiographic study. Brain Res 655:195–201

Shaw PJ, Chinnery RM, Ince PG (1994b) Non-NMDA receptors in motor neuron disease (MND): a quantitative autoradiographic study in spinal cord and motor cortex using [3H]CNQX and [3H] kainate. Brain Res 655:186–194

Shaw PJ, Ince PG, Matthews JNS, Johnson M, Candy JM (1994c) N-Methyl-D-aspartate (NMDA) receptors in the spinal cord and motor cortex in motor neuron disease: a quantitative autoradiographic study using [3H]MK-801. Brain Res 637:297–302

Shaw PJ, Forrest V, Ince PG, Richardson JP, Wastell HJ (1995a) CSF and plasma amino acid levels in motor neuron disease: elevation of CSF glutamate in a subset of patients. Neurodegeneration 4:209–216

Shaw PJ, Ince PG, Falkous G, Mantle D (1995b) Oxidative damage to protein in sporadic motor neuron disease spinal cord. Ann Neurol 38:691–695

Shaw PJ, Ince PG (1997) Glutamate, excitotoxicity and amyotrophic lateral sclerosis. J Neurol 244 [Suppl 2]:S3-S14

Shaw PJ, Chinnery RM, Thagesen H, Borthwick G, Ince PG (1997a) Immunocytochemical study of the distribution of the free radical scavenging enzymes Cu/Zn superoxide dismutase (SOD1), Mn superoxide dismutase (Mn SOD) and catalase in the normal human spinal cord and in motor neurone disease. J Neurol Sci 147:115–125

Shaw PJ, Tomkins J, Ince PG et al (1997b) Cu/Zn superoxide dismutase (SOD1) mutations in ALS CNS tissue: exclusion of somatic mutations and correlations with molecular pathology. Neurol 48 [Suppl]:A350

Siddique T, Nijhawan D, Hentat A (1996) Molecular genetic basis of familial ALS. Neurol 47 [Suppl 2]:S27-S35

Sillevis Smitt PAE, Mulder TPJ, Verspaget HW, Blaauwgeers HGT, Troost D, Vianney de Jong JMB (1994) Metallothionein in amyotrophic lateral sclerosis. Biol Signals 3(4):193–197

Subramaniam JJ, Yiannikas C (1990) Multimodality evoked potentials in motor neurone disease. Arch Neurol 47:989–994

Thatcher N, Usher PA, Shaw PJ, Ince PG (1996) The development of an immunofluorescent technique for colocalisation studies in human spinal cord using confocal microscopy. In: (eds) Proceedings of the 7th International Symposium on ALS/MND. , ,

Troy CM, Shelanski M (1994) Down regulation of copper/zinc superoxide dismutase causes apoptotic death in PC12 neuronal cells. Proc Natl Acad Sci USA 91:6384–6387

Tsai G, Stauch-Slusher B, Sim L et al (1991) Reductions in acidic amino acids and N-acetyl, aspartyl-glutamate (NAAG) in amyotrophic lateral sclerosis. Brain Res 556:151–156

Tu PH, Raju P, Robinson KA, Gurney ME, Trojanowski JQ, Lee VM (1996) Transgenic mice carrying a human mutant superoxide dismutase transgene develop neuronal cytoskeletal pathology resembling human amyotrophic lateral sclerosis lesions. Proc Natl Acad Sci USA 93:3155–3160

Volterra A, Trotti D, Racagni G (1994) Glutamate uptake is inhibited by arachidonic acid and oxygen radicals via two distinct and additive mechanisms. Mol Pharmacol 46:986–992

Wang X, Culotta VC, Klee CB (1996) Superoxide dismutase protects calcineurin from inactivation. Nature 383:434–437

Westbrook GL (1994) Glutamate receptor update. Curr Opin Neurobiol 4:337–346
White RJ, Reynolds IJ (1995) Mitochondria and Na+/Ca+ exchange buffer glutamate-induced calcium loads in cultured cortical neurons. J Neurosci 15(2):1318–1328
Wiedau-Pazos M, Goto JJ, Rabizadeh S et al (1996) Altered reactivity of superoxide dismutase in familial amyotrophic lateral sclerosis. Science 271:515–518
Williams TL, Shaw PJ, Lowe J, Bates D, Ince PG (1995) Parkinsonism in motor neuron disease: case report and literature review. Acta Neuropathol 89:275–283
Williams TL, Ince PG, Oakley AE, Shaw PJ (1996) An immunocytochemical study of the distribution of AMPA selective glutamate receptor subunits in the normal human motor system. Neuroscience 74:185–198
Williams TL, Day NC, Ince PG, Kamboj RK, Shaw PJ (1997) Calcium-permeable alpha-amino-3-hydroxy-5-methyl-4-isoxazole propionic acid receptors: a molecular determinant of selective vulnerability in amyotrophic lateral sclerosis. Ann Neurol 42:200–207
Wong PC, Pardo CA, Borchelt DR et al (1995) An adverse property of a familial ALS-linked SOD1 mutation causes motor neuron disease characterized by vacuolar degeneration of mitochondria. Neuron 14(6):1105–1116
Young AB, Penney JB (1992) Pharmacological aspects of motor dysfunction. In: Asbury AK, McKham GM, McDonald WI (eds) Diseases of the nervous system, clinical neurobiology, vol 1. Saunders, Philadelphia, pp 342–352
Young AB, Penney JB, Dauth GW, Bramberg MB, Gilman S (1983) Glutamate or aspartate as a possible neurotransmitter of cerebral corticofugal fibres in the monkey. Neurol 33:1513–1516
Zeman S, Lloyd C, Meldrum B, Leigh PN (1994) Excitatory amino acids, free radicals and the pathogenesis of motor neuron disease. Neuropathol Appl Neurobiol 20:219–231
Zhang Y, Marcillat O, Giulivi C, Ernster L, Davies KJA (1990) The oxidative inactivation of mitochondrial electron transport chain components and ATPase. J Biol Chem 265:16330–16336

5 NO–NMDA Receptor Interactions: A Neuromolecular Approach to Novel Therapeutics

S. A. Lipton, Y.-B. Choi, N. J. Sucher, and H. S.-V. Chen

5.1 Redox Modulation and NO-Related Species

In recent years, as endogenous sources of oxidizing and reducing agents have been discovered, redox modulation of protein function has been recognized as an important mechanism for many cell types. For our purposes, we will confine our review of redox modulation to covalent modification of sulfhydryl (thiol) groups on protein cysteine residues. If they possess a sufficient redox potential, oxidizing agents can react to form adducts on single sulfhydryl groups or, if two free sulfhydryl groups are vicinal (in close proximity), disulfide bonds may possibly be formed. Reducing agents can regenerate free sulfhydryl (–SH) groups by donating electron(s). Considering endogenous redox agents, in addition to the usual suspects including glutathione, ascorbate, vitamin E, lipoic acid, and reactive oxygen species, nitric oxide and its redox-related species have recently come to the fore. This has occurred largely because of the rediscovery and application to biological systems of work from the early part of this century showing the organic synthesis of nitrosothiols (RS-NO) (reviewed by Stamler et al. 1992). NO group donors represent different redox-related species of the NO group, each with its own distinctive chemistry, which lead to entirely different biological effects. NO-related species include nitric oxide ($NO^{\bullet}$) but also the other redox-related forms of the NO group: with one less electron (NO^{+}, or nitrosonium ion) or one additional electron (NO^{-}, or nitroxyl anion) (Stamler et al. 1992). Recent evidence suggests that all three of these redox-related forms or their functional equivalents are important pharmacologically and physiologically, participating in distinctive chemical reactions.

5.2 Reaction of Cysteine Sulfhydryls with the NO Group

Free endogenous nitrosonium (NO^{+}) exists only at low pH. However, functional equivalents of NO^{+} can be transferred to thiol, or more properly thiolate anion ($R\text{-}S^{-}$), at physiological pH. For example, transfer of NO^{+} equivalents occur from one nitrosothiol to another, a reaction termed transnitrosylation, i.e., $R\text{-}SH + R'\text{-}SNO \rightleftharpoons R\text{-}SNO + R'\text{-}SH$. Endogenous nitrosothiols, such as *S*-nitroso-glutathione, have been demonstrated to exist in brain and in lung and to react in this manner. The enzymatic machinery underlying the formation and breakdown of nitrosothiols is just beginning to be characterized. Recently, for example, thioredoxin reductase was shown to catalyze the homolytic cleavage of nitrosothiol (R-SNO) to nitric oxide ($NO^{\bullet} + RS^{\bullet}$) (Nikitovic and Holmgren 1996).

Very recently, the groups of Schmidt and Feelish presented evidence that neuronal nitric oxide synthase produces NO^{-} rather than NO·(Schmidt et al. 1996). NO^{-} presents an arcane chemistry since it can apparently be encountered in two different states, consisting of either a high (singlet) or low (triplet) energy state, each with distinctive chemistries. In particular, singlet NO^- can react directly with thiol while triplet NO^- does not (Bonner and Stedman 1996). However, in the triplet state NO^- may react with O_2 to form peroxynitrite ($ONOO^-$), which in turn may oxidize free thiols to disulfide (Radi et al. 1991; Kim et al. 1996).

Classically, there was no precedent for direct reaction of $NO^\bullet$ with thiols under anaerobic conditions (Pryor and Lightsey 1981; Pryor et al. 1982). Recently, however, Ischiropoulos and co-workers demonstrated that in the presence of an electron acceptor, such as O_2, $NO^\bullet$ *can* react with thiol to form a nitrosothiol (Gow et al. 1997). One perhaps overly simplistic but useful conceptualization of this reaction is that the intermediate formed by an electron acceptor and $NO^\bullet$ would be effecting NO^+ transfer in the presence of thiol. Nonetheless, the reaction of $NO^\bullet$ and superoxide anion (to form peroxynitrite would be kinetically favored over the formation of nitrosothiol if $O_2^{\bullet -}$ is also present [e.g., if $O_2^{\bullet -}$ is not scavenged by superoxide dismutase (SOD)].

Another important concept in considering the possible chemical reactions of the NO group involves our image of the local diffusion and ephemeral nature of $NO^\bullet$. Recently, Bredt and colleagues demonstrated that neuronal nitric oxide synthase (nNOS) is located very close to potential targets of NO by virtue of its PDZ domain (Brenman et al. 1996). For example, NOS interacts via its PDZ domain with the carboxyl-terminal tail of the subunit of the *N*-methyl-D-aspartate receptor (NMDAR1), that is essential for functional activity. Therefore, restricted diffusional constraints and the need for high local concentrations to facilitate NO reactions should not present a problem.

With some of the chemical reactions of these NO-related species in hand, we will next pay particular attention to the mechanism of *S*-nitrosylation or transfer of the NO moiety to cysteine sulfhydryl group(s) on the NMDA receptor. Of particular note, in recent months *S*-nitrosylation has also been shown to regulate the activity of various other ion channel proteins, G-proteins, growth factors, enzymes, and transcription factors. These reactions of NO-related species do not involve the well-known activation of guanylate cyclase by reaction with heme to increase cyclic guanosine monophosphate (cGMP) formation. Rather they involve reactions with cysteine sulfhydryls on an increasing number of protein targets to provide modulation of function, analogous to phosphorylation of critical serine, threonine, or tyrosine residues.

The chemical reactivity of the NO group and its associated redox states is related to the local redox milieu and peptide environment, pH, temperature, and the presence of catalytic amounts of transition metals. Here, we will illustrate *S*-nitrosylation (transfer of NO^+ equivalents to thiol groups) to modulate protein functional activity. The first published example of this phenomenon is represented by the reaction of the NO group with regulatory sulfhydryl(s) of the NMDA receptor's redox modulatory site(s), resulting in down-regulation of receptor activity (Lipton et al. 1993). It had been known that NO donors could decrease NMDA function (Lei et al. 1992; Manzoni et al. 1992; Manzoni and Bockaert 1993), but the exact mechanism remained in question (Fagni et al. 1995). The redox basis for this reaction will be presented below.

Most importantly, each of the NO-related species (NO^+, $NO^•$, and NO^- in singlet or triplet energy states) participates in different chemical reactions (Stamler et al. 1992; Lipton et al. 1993; Lipton and Stamler 1994). Nowhere is this more apparent than in the reactions of $NO^•$ versus NO^+ which influence neuronal survival in a diametrically opposed fashion (Lipton et al. 1993). While $NO^•$ reacts with $O_2^{•-}$ to form $ONOO^-$, which in turn triggers neurotoxic reactions either by itself or via its breakdown products, NO^+ equivalents can react with a redox modulatory site(s) on the NMDA receptor to down-regulate the receptor's activity, which produces neuroprotection (Beckman et al. 1990; Dawson et al. 1991; Lipton et al. 1993).

5.3 *S*-Nitrosylation, NMDA Receptor Down-Regulation, and Neuroprotection

The NO group can down-regulate NMDA receptor activity (Hoyt et al. 1992; Lei et al. 1992; Manzoni et al. 1992), apparently at a redox modulatory site(s) of the receptor, consisting of critical cysteine sulfhydryl or thiol group(s) (Lei et al. 1992; Lipton et al. 1993; Kohr et al. 1994; Sullivan et al. 1994). In native neurons (e.g., cerebrocortical cells), we measured amplitude of NMDA-evoked responses, monitored by whole-cell recording with a patch electrode or by digital calcium imaging with the Ca^{2+}-sensitive dye fura-2 (Lei et al. 1992; Lipton et al. 1993). We found that sulfhydryl reducing agents, such as dithiothreitol (DTT) which promote the formation of free thiol groups, increased

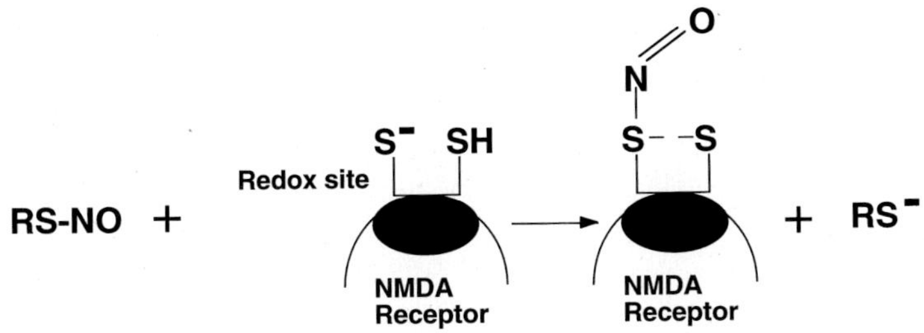

Fig. 1. *S*-nitrosylation of critical thiol group(s) of the *N*-methyl-D-aspartate receptor's (*NMDA Receptor*) redox modulatory site by *S*-nitrosocysteine (cys-NO), a more general example of which is *RS-NO*, where R is a nitrosylated protein. The redox modulatory site of the NMDA receptor is transnitrosylated by transferring the NO group (in the NO^+ form) from RS-NO to cysteine sulfhydryl group(s) on the NMDA receptor channel complex. This results in a decreased frequency of channel opening and hence decreased NMDA receptor activity

NMDA responses. In contrast, oxidizing agents, such 5,5'-dithiobisnitrobenzoic acid (DTNB) decreased NMDA responses by forming thiobenzoate protein conjugates at single sulfhydryl groups, or perhaps by facilitating disulfide bond formation. Additionally, taken together with the DTT and DTNB results, we knew that thiols on the NMDA receptor were involved because under our conditions *N*-ethylmaleimide (NEM), a relatively specific agent for alkylating thiols, irreversibly blocked the effects of these redox reagents while itself slightly decreasing responses to NMDA (Lei et al. 1992; Lipton et al. 1993). Importantly, under specific conditions, NEM also prevented the subsequent effects of NO donors, indicating that reactions of thiol and NO groups were involved. Recently, both our group and that of Bockaert (Manzoni and Bockaert 1993) have also demonstrated that endogenous production of NO can decrease NMDA receptor activity, indicating the potential physiological importance of this effect. In these experiments implicating the involvement of endogenous NO, inhibition of NOS was found to enhance subsequent NMDA receptor responses.

As an example of NO chemical reaction at the NMDA receptor, we found that *S*-nitrosocysteine (cys-NO) decreases NMDA receptor activity as demonstrated by whole-cell recording or by digital calcium imaging (Lipton et al. 1993). In the presence of SOD, cys-NO attenuated

NMDA-evoked Ca^{2+} influx, a prerequisite for NMDA receptor-mediated neurotoxicity. Not surprisingly therefore, under the same conditions, application of cys-NO ameliorated NMDA receptor-mediated neurotoxicity. These findings can be explained best by cys-NO donating NO^+ equivalents. Thus, *S*-nitrosylation or facile transfer of an NO^+ equivalent to thiol groups of the NMDA receptor (i.e., heterolytic fission of RS-NO) results in a nitrosothiol derivative of the NMDA receptor, which down-regulates receptor activity (Fig. 1). Under these conditions, any $NO^\bullet$ produced by alternative homolytic cleavage of cys-NO is prevented from entering a neurotoxic pathway of $ONOO^-$ formation via reaction with $O_2^{\bullet -}$ because of the presence of excess SOD (Lipton et al. 1993; Lipton and Stamler 1994). Rather, NO group transfer leads to down-regulation of NMDA receptor activity, possibly through facilitation of disulfide formation, although this chemical reaction has not yet been proven definitively at the NMDA receptor's redox modulatory site (Lipton et al. 1996a) (hence the *dashed line* in Fig. 1).

It is important to note that NO^+ transfer reactions may depend indirectly on catalytic amounts of transition metals. The fact that ethylenediaminetetraacetate (EDTA) can prevent the effects of the NO group on NMDA receptor activity (Fagni et al. 1995) therefore supports rather than refutes this chemistry. In particular, metals can facilitate nitrosation reactions involving $NO^\bullet$ (Stamler et al. 1992; Lipton et al. 1996). Nitrosation of redox sites is facilitated by oxygen, transition metals, and perhaps $O_2^{\bullet -}$ (Stamler 1994; Gow et al. 1997). The common event is transfer of an NO^+ equivalent or another intermediate with NO^+-like character to form an R-SNO.

In accordance with the recent report that nNOS may produce nitroxyl anion (NO^-) (Schmidt et al. 1996), we have also tested the ability of exogenous and endogenous generation of NO^- to modulate NMDA receptor function. We have found that donors of singlet NO^- can decrease NMDA receptor activity apparently via *S*-nitrosylation because the effects can be prevented by pre-treatment with NEM. Additionally, enzymatic generation of NO^-, presumably in the triplet state, can also decrease NMDA responses but apparently via formation of $ONOO^-$ following reaction of triplet NO^- with O_2 (Kim et al. 1996; Lipton et al. 1996).

Our work on nitrosylation and other redox reactions of recombinant NMDA receptors in the *Xenopus* oocyte expression system is instructive, but must also be interpreted with a degree of caution (Sullivan et al. 1994; Sucher et al. 1996). We do not yet appreciate how to form recombinant NMDA receptors that exactly mimic native receptors and therefore conclusions based on site-directed mutagenesis studies of cysteines must be viewed with tempered enthusiasm. In fact, in the course of performing polymerase chain reactions based on primers containing the cysteines known to be unique to NMDA receptor subunits, our group discovered a new NMDA receptor subunit (termed NMDAR-L or χ-1) (Ciabarra et al. 1995; Sucher et al. 1995). Additional unidentified NMDA receptor subunits probably remain to be identified. Thus, it is not yet possible to definitively understand native NMDA receptor responses based on recombinant subunits. This statement notwithstanding, our preliminary data suggest that the cysteines at position 744 and 798 are not only important to redox reactions in general, but also to the effect of NO on the NMDA receptor; however, additional cysteines on the rodent brain NR2A subunit also contribute to the NO effect. This is still very much a work in progress, but our preliminary results show that: (a) Specific NMDA receptor subunit combinations manifest larger NO-induced decreases in activity than other receptor subunit combinations (Omerovic et al. 1995; Sucher et al. 1996), and (b) at least four cysteines on NR1 and NR2A underlie the NO effect on the NMDA receptor.

5.4 Nitroglycerin Down-Regulates NMDA Receptor Activity and Ameliorates Neurotoxicity In Vitro and In Vivo

Based on the above findings, the ideal NO group donor drug would be one that reacts readily with the critical thiol group(s) of the redox modulatory site(s) of the NMDA receptor to inhibit excessive Ca^{2+} influx. We therefore studied nitroglycerin (NTG) as an exemplary compound. Specifically, this drug does not spontaneously liberate true nitric oxide ($NO^{\bullet}$) to any significant extent, and it is known to react readily with thiol groups forming derivative thionitrites (RS-NO) or thionitrates (RS-NO_2) (together, these are represented as RS-NO_x) (Lei et al. 1992; Lipton et al. 1993).

Using whole-cell recording via patch-clamp electrodes and digital calcium imaging with fura-2, we found that nitroglycerin inhibited

NMDA-evoked currents and Ca^{2+} influx (Lei et al. 1992; Lipton et al. 1993). Strong evidence that this effect of nitroglycerin is mediated by its reactions with thiol in the above-illustrated manner came from a series of chemical experiments. These studies showed that specific alkylation of thiol groups with NEM completely abrogated the inhibitory effect of nitroglycerin on subsequent NMDA-evoked responses under our conditions (Lei et al. 1992).

The finding that nitroglycerin could inhibit NMDA-evoked responses was corroborated by the demonstration that nitroglycerin also significantly ameliorates NMDA-induced neuronal killing in cerebrocortical cultures (Lei et al. 1992; Lipton et al. 1993). In addition, preliminary data suggest that high doses of nitroglycerin are neuroprotective in rat models of focal ischemia under conditions of constant systemic blood pressure and modestly increased cerebral blood flow in the penumbra (Lipton and Wang 1996). These parameters are held stable either by inducing tolerance to the systemic effects of nitroglycerin through chronic transdermal application (Sathi et al. 1993), or by intravenous infusion of a pressor agent concurrently with nitroglycerin (Lipton and Wang 1996). Although difficult to prove in vivo, it appears likely that the decrease in stroke size observed after treatment with nitroglycerin is at least in part due to its effect on decreasing NMDA receptor activity, although other beneficial actions are also possible (Lipton and Wang 1996).

5.5 *S*-Nitrosylation of Cysteine Sulfhydryls on Other Ion Channels, Enzymes, Transcription Factors, and Regulatory Proteins

Shortly after NMDA receptor activity was shown to be regulated by NO-related species, similar data were presented for the Ca^{2+}-activated K^+ channel of cardiac muscle (Bolotina et al. 1994). In this case, donors of NO^+ equivalents were shown to activate the channel, and, similar to findings at the NMDA receptor in our laboratory, NEM blocked the effect by irreversibly alkylating thiol groups.

Along these lines, several other ion channels, enzymes, G-proteins, transcription factors, and other proteins are either up-regulated or down-regulated by similar mechanisms of *S*-nitrosylation or donation of NO^+

equivalents to regulatory sulfhydryl centers (Stamler et al. 1997). The list will undoubtedly grow just as in recent years phosphorylation, myristolation, and palmitoylation have become recognized as important biochemical processes for regulatory function. Interestingly, palmitoylation may be aimed at similar critical thiol group targets, resulting in thioester bond formation. In fact, on some proteins such as synaptosomal-associated protein (SNAP)-25 it is possible that *S*-nitrosylation and palmitoylation may compete for the same sulfhydryl, possibly with different physiological outcomes (Hess et al. 1993).

In contrast to phosphorylation, however, in the case of *S*-nitrosylation evidence is accumulating that the critical cysteine residues may be located extracellularly, intracellularly, or possibly even within the putative membrane spanning region of a protein. From this point of view *S*-nitrosylation may offer additional versatility in the modes of control that can be exerted compared to phosphorylation and other better known post-translational forms of modification.

5.6 A Candidate Consensus Motif for *S*-Nitrosylation

Many functionally important sites or target sites for post-translational modification of proteins are distinguished in the primary amino acid sequence by the occurrence of certain patterns or motifs. In many cases, such motifs constitute only a very minor part of the entire protein primary sequence. Thus, small patterns often are not detected by overall alignment of protein sequences that are only distantly or not at all related. Such motifs, however, can be identified by the occurrence of a particular cluster of residue types in the primary sequence. A collection of such sequence fingerprints has been developed for PROSITE, a database of biologically significant sites and patterns that can be used to identify families of functionally related proteins or sites for post-translational modification. Examples of such motifs are the consensus sequence patterns required for glycosylation or phosphorylation.

In an attempt to define a possible consensus motif that might be required, or at least be facilitatory for *S*-nitrosylation, we initially examined the putative target sites for redox modulation of NMDA receptors. Most importantly, cysteine residues in similar motifs to that described below for the NMDA receptor have been shown by various chemical criteria to be nitrosylated on proteins such as hemoglobin, p21^{ras}, cyclooxygenase, and others (Stamler et al. 1997). Two cysteines (abbreviated as C in the single-letter amino acid code) in the NMDAR1 subunit have been found by site-directed mutagenesis to be necessary for redox modulation of that receptor (Sullivan et al. 1994). Unexpectedly, however, these cysteines, C744 and C798, appear to be conserved in all ionotropic glutamate receptors when the sequences are aligned by overall homology (Moriyoshi et al. 1991; Sucher et al. 1995). Nonetheless, among the ionotropic glutamate receptors, only NMDA receptors are exquisitely sensitive to redox modulation (Aizenman et al. 1989). Inspection of the immediate amino acid neighbors of these cysteines revealed that the NMDAR1 cysteines are distinguished from the cysteines conserved in the other ionotropic glutamate receptors in that they are preceded at position –2 by a polar amino acid (G,S,T,C,Y,N,Q), an acidic (D,E) or basic (K,R,H) amino acid at position –1, and an acidic amino acid at position +1. Based on this observation, we constructed the degenerate amino acid pattern designated (G,S,T,C,Y,N,Q)(K,R,H,D,E)C(D,E) in standard single-letter amino acid code and used it in a search of the Protein Identification Resource (PIR) and Swiss Protein (SW) databases with the program Findpatterns of the GCG software package (Program Manual for the Wisconsin Package, Version 8, September 1994, Genetics Computer Group, 575 Science Drive, Madison, 53711 WI, USA). In the PIR database (Release 44.0; March 1995), 3878 sequences out of 77 573 contained this pattern at least once, while in the SW database (Release 31.0; March 1995), 2383 out of 43 470 sequences contained this pattern. Viral, bacterial, plant, and animal sequences contained this motif.

Among candidates for regulation by *S*-nitrosylation that were identified by the database search were ion channels (NMDA receptor, voltage-sensitive Na^+ channel, cyclic nucleotide-gated channel), transporters (Ca-ATPase, K-transporter), receptors (inositol trisphosphate receptor, nerve growth factor receptor), enzymes (oxidoreductases, dehydrogenases, adenylate and guanylate cyclases, proteases, DNA topoisomerases, DNA and RNA polymerases, kinases, phosphatases), transcription factors [helix loop helix proteins, nuclear factor (NF)-κB, zinc finger proteins], small GTP binding proteins (rab, ras, sas, ypt), cell adhesion molecules (integrins, neural cell adhesion molecule), cell adhesion substrates (laminin, collagen), cyclins and coagulation factors (IXa, Xa, XIII).

In fact, 20 out of 27 proteins that had been listed in a recent review (Stamler 1994) as bioregulatory targets of nitrogen oxides contain the full motif. The presence of the putative "nitrosylation" motif in guanylate cyclase suggests that the NO group may regulate the functional activity of guanylate cyclase by *S*-nitrosylation in addition to the interaction with the heme group of this enzyme. It is possible, however, that *S*-nitrosylation might occur at sites other than the proposed motif, or that the motif may only be evident in the tertiary rather than the primary structure of some proteins. Moreover, it appears that a certain subset of the motif may bear the highest statistical correlation to the propensity for nitrosylation (Stamler et al. 1997). While the proposed motif was defined post hoc based on our results with redox modulation of the NMDA receptor, it should allow us to identify possible target proteins for *S*-nitrosylation. Most importantly, this motif predicts a target sequence that can be subjected to site-directed mutagenesis in order to experimentally verify its importance for *S*-nitrosylation.

5.7 Conclusions

In summary, the possible chemical reactions of the NO group are dictated by its redox state. In the case of NO^+ equivalents, this mechanism appears to involve *S*-nitrosylation and possibly further oxidation of critical thiols to disulfide bonds in the NMDA receptor's redox modulatory site(s) to down-regulate channel activity. Very recent data also suggest that NO^-, probably in the singlet state, can react with critical sulfhydryl group(s) of the NMDA receptor to down-regulate its activity; in the triplet state NO^- may oxidize these NMDA receptor sulfhydryl groups by formation of an intermediate such as $ONOO^-$ (Lipton et al. 1996). $NO^\bullet$ can react with thiol if an electron acceptor such as O_2 is present and if the conditions do not favor the kinetically preferred reaction with $O_2^{\bullet-}$ to yield $ONOO^-$.

It is becoming increasingly evident that in addition to NMDA receptors, biological activities of many other proteins containing critical cysteine residues can be regulated by *S*-nitrosylation and other redox reactions in a sense similar to the type of control exerted by phosphorylation (Lipton et al. 1993). This type of chemical reaction may represent a new and ubiquitous pathway for the molecular control of protein function by potentially reactive sulfhydryl centers.

Acknowledgements. This review is based upon work performed in close collaboration between the laboratories of Stuart Lipton at Harvard Medical School and Jonathan Stamler at Duke Medical Center. We are grateful to the many members of the two laboratories who contributed to the work. This was a case in which the mutual ignorances of the two groups exactly complemented one another, resulting in a true collaboration. The work was presented nearly simultaneously at three meetings – in Kyoto, Berlin, and Barcelona – and therefore three related versions of this manuscript were offered for the proceedings of these meetings.

References

Aizenman E, Lipton SA, Loring RH (1989) Selective modulation of NMDA responses by reduction and oxidation. Neuron 2:1257–1263

Arnelle DR, Stamler JS (1995) NO^+, $NO^\bullet$, and NO^- donation by *S*-nitrosothiols: implications for regulation of physiological functions by *S*-nitrosylation and acceleration of disulfide formation. Arch Biochem Biophys 318:279–285

Beckman JS, Beckman TW, Chen J, Marshall PA, Freeman BA (1990) Apparent hydroxyl radical production by peroxynitrite: implications for endothelial injury from nitric oxide and superoxide. Proc Natl Acad Sci USA 87:1620–1624

Bolotina VM, Najibi S, Palacino JJ, Pagaon PJ, Cohen RA (1994) Nitric oxide directly activates calcium-dependent potassium channels in vascular smooth muscle. Nature 368:850–853

Bonner FT, Stedman G (1996) The chemistry of nitric oxide and redox-related species. In: Feelisch M, Stamler JS (eds) Methods in nitric oxide research. Wiley, Chichester, England

Brenman JE, Chao DS, Gee SH, McGee AW, Craven SE, Santilliano DR, Wu Z, Huang F, Xia H, Peters MF, Froehner S, Bredt DS (1996) Interaction of nitric oxide synthase with the postdynaptic density protein PSD-95 and α1-syntrophin mediated by PDZ domains. Cell 84:757–767

Ciabarra AM, Sullivan JM, Gahn LG, Pecht G, Heinemann S, Sevarino KA (1995) Cloning and characterization of χ–1: a developmentally regulated member of a novel class of the ionotropic gutamate receptor family. J Neurosci 15:6498–6508

Dawson VL, Dawson TM, London ED, Bredt DS, Snyder SH (1991) Nitric oxide mediates glutamate neurotoxicity in primary cortical cultures. Proc Natl Acad Sci USA 88:6368–6371

Fagni L, Olivier M, Lafon-Cazal M, Bockaert J (1995) Involvement of divalent ions in the nitric oxide-induced blockade of N-methyl-D-aspartate receptors in cerebellar granule cells. Mol Pharmacol 47:1239–1247

Gow AJ, Buerk DG, Ischiropoulos H (1997) A novel reaction mechanism for the formation of S-nitrosothiol in vivo. J Biol Chem 272:2841–2845

Hess DT, Patterson SI, Smith DS, Skene JHP (1993) Neuronal growth cone collapse and inhibition of protein fatty acylation by nitric oxide. Nature 366:562–565

Hoyt KR, Tang L-H, Aizenman E, Reynolds IJ (1992) Nitric oxide modulates NMDA-induced increases in intracellular Ca^{2+} in cultured rat forebrain neurons. Brain Res 592:310–316

Kim W-K, Rayudu PV, Mullins ME, Stamler JS, Lipton SA (1996) Down regulation of NMDA receptor activity in cortical neurons by peroxynitrite. In: Moncada S, Stamler JS, Gross S, Higgs EA (eds) The biology of nitric oxide, part 5. Portland, London

Kohr G, Eckardt S, Lüddens H, Monyer H, Seeburg PH (1994) NMDA receptor channels: subunit-specific potentiation by reducing agents. Neuron 12:1031–1040

Lei SZ, Pan Z-H, Aggarwal SK, Chen H-SV, Hartman J, Sucher NJ, Lipton SA (1992) Effect of nitric oxide production on the redox modulatory site of the NMDA receptor-channel complex. Neuron 8:1087–1099

Lipton SA, Stamler JS (1994) Actions of redox-related congeners of nitric oxide at the NMDA receptor. Neuropharmacology 33:1229–1233

Lipton SA, Wang YF (1996) NO-related species can protect from focal cerebral ischemia/reperfusion. In: Krieglstein J (ed) Pharmacology of cerebral ishcemia. Medpharm Scientific, Stuttgart, pp 183–191

Lipton SA, Choi Y-B, Pan Z-H, Lei SZ, Chen H-SV, Sucher NJ, Loscalzo J, Singel DJ, Stamler JS (1993) A redox-based mechanism for the neuroprotective and neurodestructive effects of nitric oxide and related nitroso-compounds. Nature 364:626–632

Lipton SA, Choi Y-B, Sucher NJ, Pan Z-H, Stamler JS (1996a) Redox state, NMDA receptors, and NO-related species. Trends Pharmacol Sci 17:186–187

Lipton SA, Kim W-K, Rayudu PV, Asaad W, Arnelle DR, Stamler JS (1996b) Singlet and triplet nitroxyl anion (NO^-) lead to N-methyl-D-aspartate (NMDA) receptor downregulation and neuroprotection. In: Stamler Gross JS, Moncada S (eds) The biology of nitric oxide. Portland, London

Manzoni O, Bockaert J (1993) Nitric oxide synthase activity endogenously modulates NMDA receptors. J Neurochem 61:368–370

Manzoni O, Prezeau L, Marin P, Deshager S, Bockaert J, Fagni L (1992) Nitric oxide-induced blockade of NMDA receptors. Neuron 8:653–662
Moriyoshi K, Masu M, Ishii T, Shigemoto R, Mizuno N, Nakanishi S (1991) Molecular cloning and characterization of the rat NMDA receptor. Nature 354:31–37
Nikitovic D, Holmgren A (1996) S-nitrosoglutathione is cleaved by the thioredoxin system with liberation of glutathione and redox regulating nitric oxide. J Biol Chem 271:19180–19185
Omerovic A, Chen S-J, Leonard JP, Kelso SR (1995) Subunit-specific redox modulation of NMDA receptros expressed in Xenopus oocytes. J Recept Sign Transduc Res 15:811–827
Pryor WA, Lightsey JW (1981) Mechanisms of nitrogen dioxide reactions: initiation of lipid peroxidation and the production of nitrous acid. Science 214:435–437
Pryor WA, Church DF, Govinden CK, Crank G (1982) Oxidation of thiols by nitric oxide and nitrogen dioxide: synthetic utility and toxicological implications. J Org Chem 47:156–159
Radi R, Beckman JS, Bush KM, Freeman BA (1991) Peroxynitrite oxidation of sulfhydryls. The cytotoxic potential of superoxide and nitric oxide. J Biol Chem 266:4244–4250
Sathi S, Edgecomb P, Warach S, Manchester K, Donaghey T, Stieg PE, Jensen FE, Lipton SA (1993) Chronic transdermal nitroglycerin (NTG) is neuroprotective in experimental rodent stroke models. Soc Neurosci Abstr 19:849
Schmidt HHHW, Holman H, Schindler U, Shutenko ZS, Cunningham DD, Feelisch M (1996) No $^{\bullet}$NO from NO synthase. Proc Natl Acad Sci USA 93:14492–14497
Stamler JS (1994) Redox signaling: nitrosylation and related target interactions of nitric oxide. Cell 78:931–936
Stamler JS, Singel DJ, Loscalzo J (1992) Biochemistry of nitric oxide and its redox activated forms. Science 258:1898–1902
Stamler JS, Toone EJ, Lipton SA, Sucher NJ (1997) (S)NO signals: translocation, regulation, and a consensus motif. Neuron 18:691–696
Sucher NJ, Schahram A, Chi CL, Leclerc CL, Awobuluyi M, Deitcher DL, Wu MK, Yuan JP, Jones EG, Lipton SA (1995) Developmental and regional expression pattern of a novel NMDA receptor-like subunit (NMDAR-L) in the rodent brain. J Neurosci 15:6509–6520
Sucher NJ, Awobuluyi M, Choi Y-B, Lipton SA (1996) NMDA receptors: from genes to channels. Trends Pharmacol Sci 17:348–355
Sullivan JM, Traynelis SF, Chen H-SV, Escobar W, Heinemann SF, Lipton SA (1994) Identification of two cysteine residues that are required for redox modulation of the NMDA subtype of glutamate receptor. Neuron 13:929–936

6 Glutamate in Neurodegenerative Disorders: Phenotype Resulting from Decreased Expression of AMPA Receptor GluR-B Subunit mRNA in Rats

G. Müller, U. Endermann, I. Bresink, and L. Turski

6.1 Introduction

The excitatory neurotransmitter glutamate activates various classes of ionotropic receptors in the mammalian central nervous system. α-Amino-3-hydroxy-5-methyl-4-isoxazolepropionate (AMPA) and kainate (KA) receptor stimulation triggers Na^{+} influx into neurons, while *N*-methyl-D-aspartate (NMDA) receptor stimulation additionally leads to considerable Ca^{2+} influx (Hollmann and Heinemann 1994). Relative impermeability of the AMPA receptor channel to Ca^{2+} is regulated by the Q/R locus within the channel pore on the M2 region of the GluR-B subunit. Expression of either glutamine (Q) or arginine (R) within this locus is controlled by posttranscriptional RNA editing (Sommer et al. 1991). Transgenic mice engineered to synthesize unedited GluR-B subunits express Ca^{2+}-permeable AMPA receptors. These mice develop progressive neurodegeneration and spontaneous seizures (Brusa et al. 1995).

The mutant strain of Wistar rats *spa/spa* develops a progressive neurodegenerative syndrome which follows an autosomal recessive inheritance pattern. These animals have lower GluR-B subunit mRNA expression in the cerebellum and hippocampus than littermate controls (Margulies et al. 1993) and express Ca^{2+}-permeable AMPA receptors. Life expectancy of *spa/spa* rats is 8–10 weeks. They develop motor deficits and progressive neuronal degeneration in the CNS (Wagemann et al. 1991). Due to the longer life expectancy of *spa/spa* rats than GluR-B transgenic mice these animals allow evaluation of the consequences of decreased GluR-B subunit expression on the phenotype.

To address the question of whether enhanced Ca^{2+} permeability of AMPA channels causes phenotypic sequelae we evaluated growth, skeletal anomalies, motor disturbances, gait pattern, and exploratory activity in *spa/spa* rats. To address the issue of neuronal degeneration in the CNS of these animals, we analyzed brains and spinal cords morphometrically. Furthermore, we employed long-term treatment with glutamate receptor antagonists in an attempt to elucidate whether they are able to prevent neuronal death triggered by changes in GluR-B subunit expression in *spa/spa* rats.

6.2 Phenotype in Genetically Determined Rats with Decreased Expression of GluR-B Subunit mRNA in the Brain

6.2.1 Growth and Skeletal Anomalies

6.2.1.1 Body Weight

Male *spa/spa* rats (Charles River, Haarlan, Holland), littermates, and wildtype rats (Wistar) were housed under environmentally controlled conditions (6:00 A.M.–6:00 P.M.; 12-h light/dark cycle; 22°–24°C; 40%–60% humidity) and permitted free access to food and water. Body weight was monitored by means of a Sartorius model U6100 balance starting at the mean age of 39.2±1.1 days. The measurements were performed daily between 8:00 and 9:00 A.M. and were initiated as early as phenotypic differences between *spa/spa* and littermates could be recognized.

Figure 1A shows that between 30–50 days body weight in *spa/spa* rats did not significantly differ from that in littermates, both ranging between 80 and 120 g. Between day 50 and 80 the littermates doubled their body weight while in *spa/spa* rats there was no increment of body weight, which remained between 100 and 120 g and showed a tendency to drop during the last 10 days of life.

6.2.1.2 Body Length

Starting at an average age of 41.6±1.1 days, rat body length was determined by means of a metric scale during maximal stretching. The distance between the tip of the tail and the nose was defined as the body length. Measurements were performed twice a week between 9:00 and 12:00 A.M.

Figure 1B shows that in *spa/spa* rats body length increased from 160 to 180 mm from day 30 to 50 and did not differ from that in littermates. During the following 30 days body length of littermates increased to 220 mm, whereas *spa/spa* rats did not grow.

6.2.1.3 Skeletal Anomalies

Beginning at an average age of 35±0.6 days, length and width of the skull and the degree of thoracic kyphosis were measured by means of

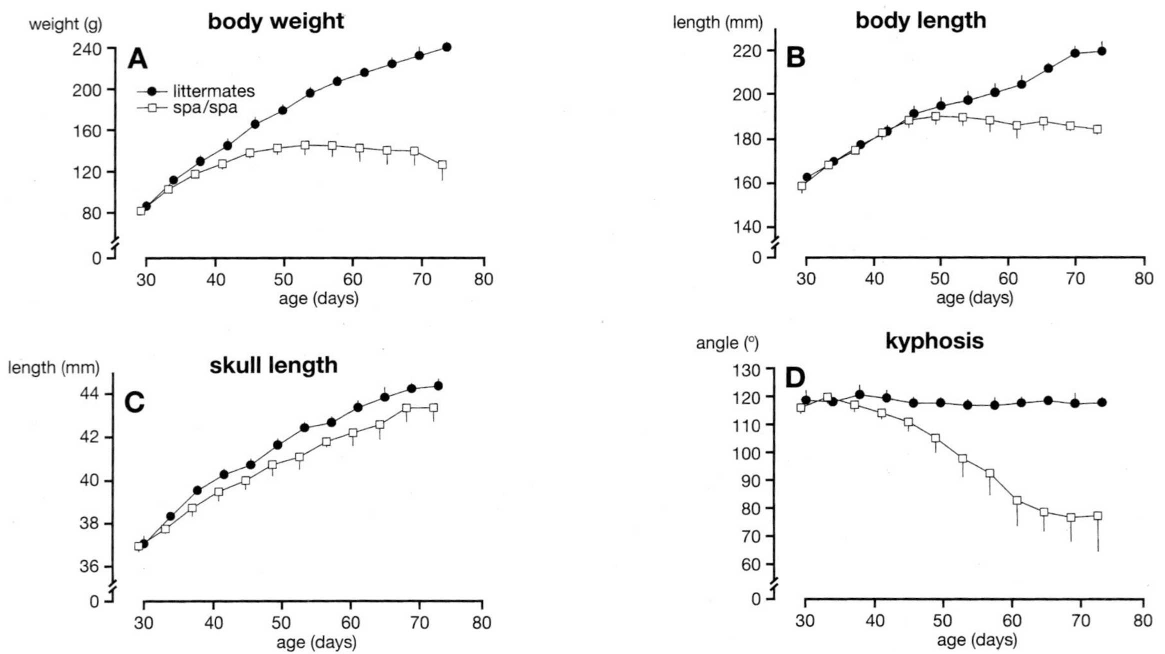

Fig. 1A–D. Growth and skeletal anomalies in genetically determined *spa/spa* rats. Body weight (**A**) was monitored by means of a balance scale in groups of 13 *spa/spa* rats and 13 littermates. Body length (**B**) was determined by means of a metric scale in groups of 10 *spa/spa* rats and 11 littermates. Length of the skull (**C**) and the degree of thoracic kyphosis (**D**) were measured by X-ray radiographs

X-ray radiographs taken during ketamine/xylazine anesthesia. Measurements were performed twice a week between 1:00 and 4:00 P.M. by means of a precision caliper (Digi-Met; Preisser, Meiningen, Germany).

Skull length increased in littermates from 37.1±0.4 mm to 44.1±0.2 mm, whereas in *spa/spa* rats it increased from 37.0±0.3 mm to 43.2±0.6 mm (Fig. 1C). Skull width had a similar developmental profile and progressed from 17.0 mm to 19.4±0.4 mm in littermates and from 16.7±0.1 mm to 18.8±0.6 mm in *spa/spa* rats.

Kyphosis severity was assessed by measuring the angle of cervical lordosis between vertebrae C2 and Th3 using a precision goniometer. Littermates did not show evidence of kyphosis. The angle remained between 118°±4° and 116°±4° during the entire observation period of up to 70 days. In *spa/spa* rats kyphosis rapidly developed following day 40 and reached 75°±11° by the age of 72 days (Fig. 1D).

The analysis of growth and skeletal anomalies showed that *spa/spa* rats did not differ from littermates up to the age of 30–40 days. After the age of 40–50 days *spa/spa* rats grew less rapidly and developed kyphosis.

6.2.2 Motor Disturbances and Gait Patterns

6.2.2.1 Motor Disturbances

Starting at an average age of 42 days, righting reflex, hindlimb dragging, muscle tone, and tremor were assessed in an open field (60×45 cm) for at least 3 min between 9:00 and 12:00 A.M. The inability of rats to regain an upright position within 30 s was scored as a loss of righting reflex. Hindlimb dragging was scored when rats could move forward holding their hindlimbs in extension. Muscle tone was rated as increased when abdominal muscles were resistant to palpation. Tremor was defined as rhythmic whole-body trembling for longer than 30 s. Littermates regained their righting reflex within a few seconds. In contrast, by the age of 70–74 days 75% of *spa/spa* rats had lost their righting reflex (Fig. 2A).

Hindlimb dragging became evident in *spa/spa* rats between 40–50 days of age and was seen in 100% of animals at the age of 70 days (Fig. 2B). Increased muscle tone was present as early as the age of 30–40 days in 30%–40% of *spa/spa* rats.

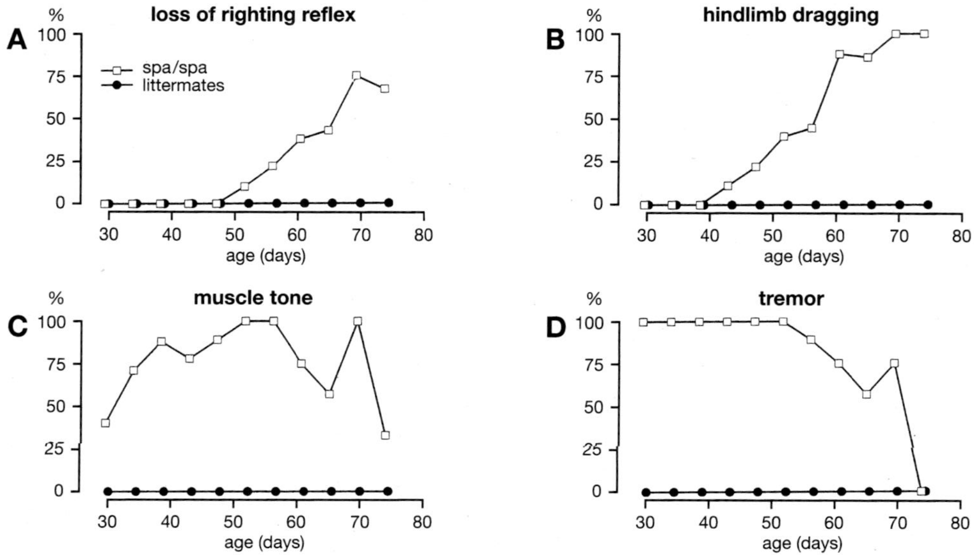

Fig. 2A–D. Motor disturbances in genetically determined *spa/spa* rats. Loss of righting reflex (**A**; in groups of 10 *spa/spa* rats and 11 littermates), hindlimb dragging (**B**), muscle tone (**C**), and tremor (**D**) were assessed in an open field (60×45 cm)

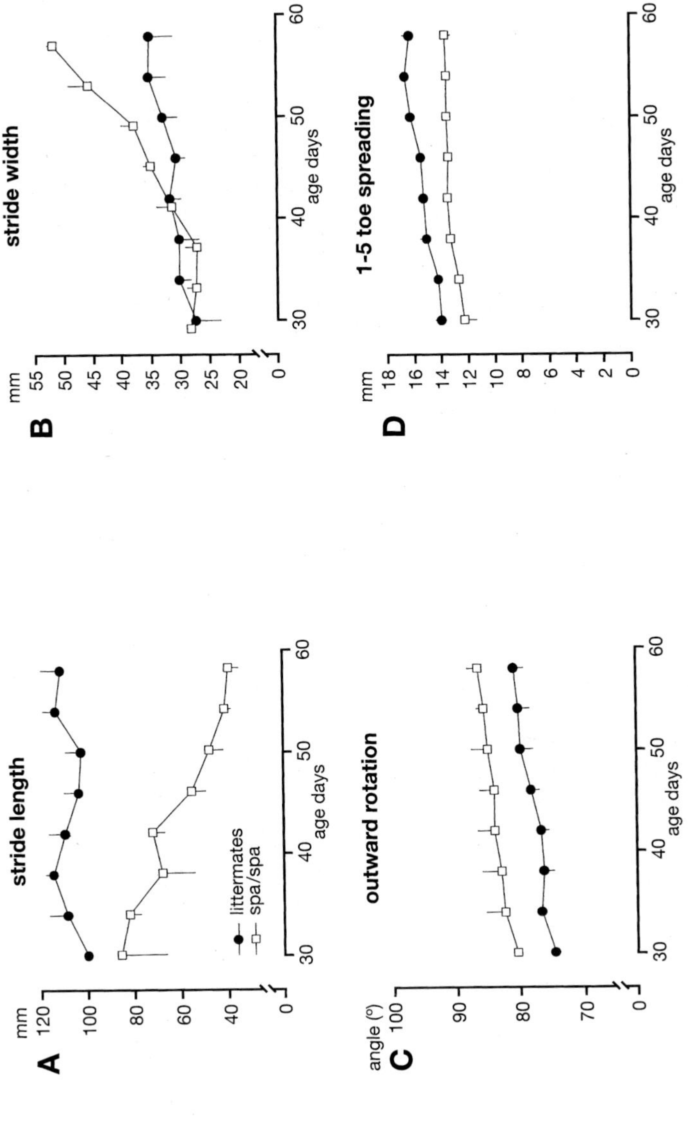

Fig. 3A–D. Gait patterns in genetically determined *spa/spa* rats. Stride length (**A**; in groups of 8 *spa/spa* rats and 9 littermates), width (**B**), outward rotation angle (**C**) and 1–5 toe spreading (**D**) are given

It progressed rapidly in severity within the next 10 days and remained severe until death between 70 and 80 days of age (Fig. 2C). Tremor in *spa/spa* rats had appeared by the age of 34 days and decreased shortly before death (Fig. 2D).

6.2.2.2 Gait Patterns

Beginning at the age of 42 days stride length, width, 1–5 toe spreading, and outward rotation angle were assessed twice a week between 11:00 A.M. and 4:00 P.M. The gait analysis was performed according to the methods of Clarke and Parker (1988) and Gurney et al. (1994).

The rat had to walk through an 80 cm long and 10 cm wide tunnel which had 30-cm-high black walls and was equipped with a Perspex bottom for videotaping. The tunnel was placed in a sound-proof room. The legs of each rat were stained with washable ink. Up to eight steps were then made by the freely moving animal. The steps were videotaped, digitized, and subsequently analyzed (Image 1.51, NIH, Bethesda, USA). Stride length in littermates remained constant between the ages of 30 and 58 days (Fig. 3A), varying from 108.3±7.9 mm at the age of 30 days to 111.1±8.6 mm in 58-day-old rats (Fig. 3A). In *spa/spa* rats the stride length progressively decreased from 81.7±4.7 mm at the age of 30 days to 37.7±4.4 mm at the age of 58 days (Fig. 3A).

In littermates between 30–58 days no changes in stride width were detected, varying from 30.3±2.0 mm to 36.6±4.0 mm. *spa/spa* rats demonstrated an increase in step width from 27.5±1.6 mm to 51.6±0.9 mm over time (Fig. 3B).

The outward rotation angle in littermates increased from 76.6°±0.7° at 30 days to 78.5°±2.1° at 58 days, whereas in *spa/spa* rats it increased from 82.3°±3° at 30 days to 86°±1.6° at 58 days (Fig. 3C).

The spreading between the 1st and 5th toe increased in littermates from 14.3±0.1 mm at 30 days to 16.9±0.3 mm at 58 days, whereas in *spa/spa* rats it increased from 12.8±0.3 mm to 13.5±0.4 mm (Fig. 3D).

Gait analysis showed that *spa/spa* rats shortened and widened their strides, increased outward rotation angle of their legs, and decreased 1–5 toe spreading in an attempt to compensate for motor coordination disturbances appearing at the age of 30–40 days.

6.2.3 Exploratory Activity

A computerized Digiscan-16 Animal Activity Monitoring System (Omnitech, Columbus, OH, USA) was used to quantify and track locomotor activity in rats. Each activity monitor consisted of a Perspex box (40×40×40 cm) surrounded by 48 horizontal and vertical infrared sensor beams. The total number of interruptions of the horizontal sensors was taken as a measure of horizontal activity, while that of vertical sensors was taken as a measure of vertical activity (Klockgether and Turski 1990). The total distance traveled was measured in centimeters and average speed of movements was measured in centimeters/second. The time spent by rats moving in close proximity to the walls (closer than 1 cm) or in the center (>1 cm) was also recorded. Starting at the age of 40 days locomotor activity monitoring and movement tracking was performed on independent groups of rats twice a week for 2 min between 9:00 and 11:00 A.M. Rats showing a pattern of activity limited to movements along the cage walls or in one cage compartment (i.e., avoiding the center of the cage) were designated "anxious." Therefore the center time/margin time ratio was used as a measure of anxiety (Pinna et al. 1997).

Horizontal activity in littermates ranged from 800 to 1000 counts/2 min between 30 and 74 days of age. In *spa/spa* rats horizontal activity decreased from 1002±91 counts/2 min at day 42 to 79±24 counts/2 min at day 74 (Fig. 4A). Similarly, the total distance, average speed, and vertical activity decreased in *spa/spa* rats progressively, whereas it remained unchanged in littermates between 30 days and 74 days (Fig. 4B–D).

The center time/margin time ratio remained very low in *spa/spa* rats up to the age of 8 weeks (in groups of 9–11). In contrast, littermates showed progressive increase in the ratio between 4 and 8 weeks (in groups of 9–13).

These observations suggest that exploratory activity in *spa/spa* rats was progressively restricted beginning from day 30–40 and ceased entirely by the age of 70–80 days. Low center time/margin time ratio indicates anxiety in *spa/spa* rats.

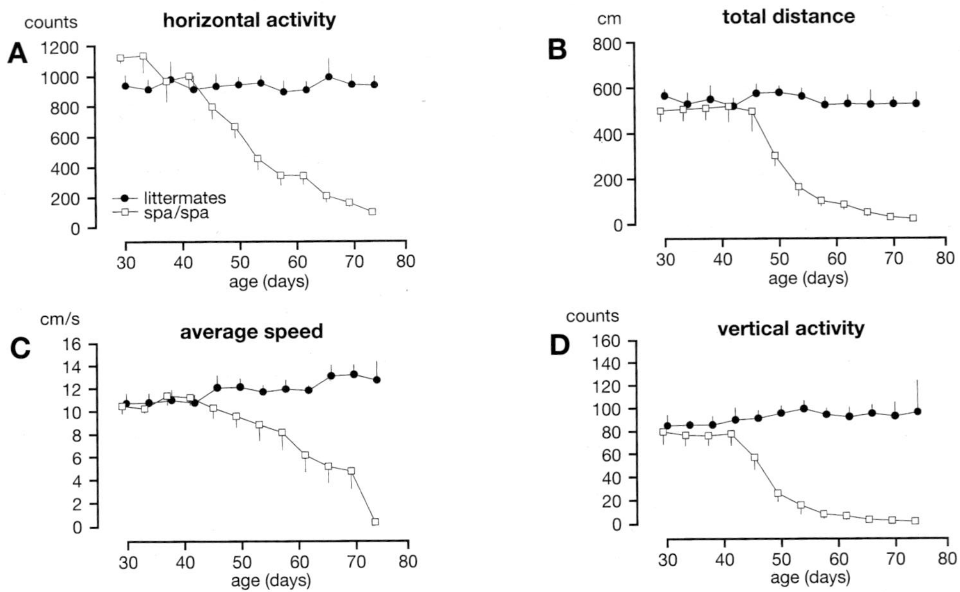

Fig. 4A–D. Exploratory activity in genetically determined *spa/spa* rats. Horizontal activity (**A**), total distance (**B**), average speed (**C**) and vertical activity (**D**) were quantified by means of a computerized Digiscan-16 Animal Activity Monitoring System in groups of 13 *spa/spa* rats and 13 littermates

Table 1. Regional distribution of (*S*)-[^{3}H]fluorowillardiine binding in rat brain as determined by quantitative autoradiography

	4 weeks		*9 weeks*	
	wt	*spa/spa*	wt	*spa/spa*
Dentate gyrus	257±14 (9)	246±12 (10)	248±8 (6)	212±14 (6)
CA1	258±12 (9)	240±12 (10)	241±4 (6)	222±9 (6)
CA3	211±14 (9)	201±14 (10)	184±6 (6)	157±9* (6)

Optical densities were converted to fmol/mg wet weight using the tritium microscales. Nonspecific binding was determined with 1 m*M* L-glutamate and subtracted from all readings. Values are the means±SEM of readings from at least six rats. Number of rats are in parentheses. *$p<0.05$ vs. 4-week-old *spa/spa* rats (Student's *t* test).

6.2.4 Autoradiography

For determination of AMPA receptor distribution with autoradiography (Hawkins et al. 1995) [^{3}H]fluorowillardiine was used as radioligand. *spa/spa* and wildtype rats were killed by decapitation, and their brains were removed rapidly and frozen in isopentane cooled with liquid nitrogen. Serial cryostat sections of 10 µm thickness were thaw-mounted on gelatine-coated slides and stored at –80°C. The sections were preincubated (twice for 15 min) in 50 m*M* Tris-HCl buffer (pH 7.4 at 4°C). The sections were then incubated in 50 m*M* Tris-HCl buffer containing 10 nmol/l (*S*)-[^{3}H]fluorowillardiine (Tocris, Buckhurst Hill, UK) for 60 min at 4°C and washed in 50 m*M* Tris-HCl buffer (3×) and water (2×) at 4°C. Finally, the sections were dried with cool air. Nonspecific binding was determined in the presence of 1 m*M* L-glutamate. The sections and [^{3}H]-labeled microscales (Amersham) were exposed to [^{3}H]Hyperfilm (Amersham) for 4 weeks. Subsequently the films were developed and quantified by computer-assisted densitometry. Three to five determinations were made for each region analyzed.

Quantitative autoradiography with [^{3}H]fluorowillardiine revealed that the density of AMPA receptors within the hippocampal subfield CA3 was lower in 9-week-old *spa/spa* rats than the AMPA receptor density in 9-week-old wildtype rats. No such changes were seen in 4-week-old rats (Table 1, Fig. 5). A decrease in [^{3}H]fluorowillardiine binding was not visible in the subfield CA1 and in the dentate gyrus

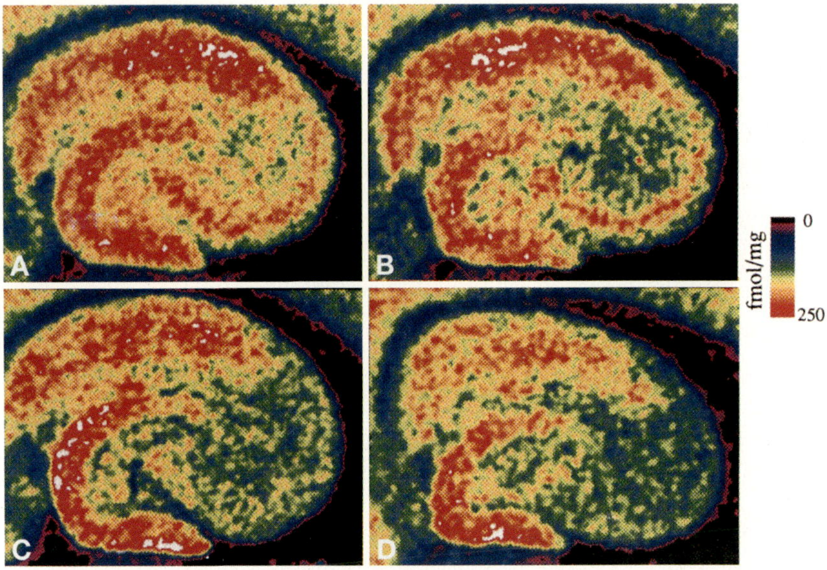

Fig. 5A–D. Autoradiography of [^{3}H]fluorowillardiine in the hippocampus of wildtype (**A**, 4-week-old; **B**, 9-week-old) and *spa/spa* (**C**, 4-week-old; **D**, 9-week-old) rats. For determination of AMPA receptor distribution in the hippocampus [^{3}H]fluorowillardiine was used. Brain sections were incubated for 60 min in 50 m*M* Tris-HCl buffer containing 10 nmol/l (*S*)-[^{3}H]fluorowillardiine. Nonspecific binding was determined in the presence of 1 m*M* L-glutamate

(Fig. 5). Interestingly, the number of AMPA receptors in the subfield CA3 decreased with increasing age in wildtype rats, as it did in *spa/spa* rats (Fig. 5). The relative decrease in the density of AMPA receptors was more pronounced in *spa/spa* rats (Fig. 5).

6.2.5 Neuropathology

For morphological examination by light microscopy *spa/spa* and wildtype rats were anesthetized with an overdose of pentobarbital (Ceva; Neuilly-sur-Seine, France) and perfused with saline, followed by fixative containing 4% paraformaldehyde and 10% 1 *M* sodium phosphate. Brains were allowed to fix in situ for 24 h and were then removed and processed for paraffin embedding. Serial coronal sections of whole brain and transversal sections of the lumbar region of spinal cord were cut at 10 μm thickness, and every 20th section was mounted on glass slides and stained with cresyl violet. To assess neuronal loss in the hippocampus, cerebellum, and spinal cord quantitatively an unbiased stereological disector technique (Cruz-Orive and Weibel 1990) was used to estimate the mean numerical density (N_v) of pyramidal neurons, Purkinje cells, or α-motoneurons. The N_v for each brain or spinal cord region was determined with 5–10 dissectors per animal.

The density of pyramidal cells in the hippocampal subfield CA3 in wildtype rats at the age of 9 weeks was $0.189 \pm 0.006 \times 10^6/mm^3$ ($n=8$). In *spa/spa* rats the density of neurons in the hippocampal CA3 subfield was decreased by 38%, resulting in an absolute value of $0.117 \pm 0.005 \times 10^6/mm^3$ ($n=15$).

In the subfield CA1, densities of pyramidal cells did not differ between wildtype and *spa/spa* rats ($0.200 \pm 0.004 \times 10^6/mm^3$ vs. $0.199 \pm 0.004 \times 10^6/mm^3$). In the cerebellum the density of Purkinje cells in wildtype rats was $0.029 \pm 0.001 \times 10^6/mm^3$, whereas in *spa/spa* rats it was decreased by 38%, resulting in an absolute value of $0.018 \pm 0.001 \times 10^6/mm^3$. The density of α-motoneurons in wildtype rats was $0.0045 \pm 0.0002 \times 10^6/mm^3$, whereas that in *spa/spa* rats was $0.0035 \pm 0.0002 \times 10^6/mm^3$ (23% decrease).

These observations demonstrate that at the age of 9 weeks *spa/spa* rats show neuronal loss in the hippocampal subfield CA3 but not in CA1, and loss of cerebellar Purkinje neurons and α-motoneurons in the spinal cord.

6.2.6 Therapy with Glutamate Antagonists

To provide continuous administration of the AMPA antagonist 7-morpholino-2,3-dioxo-6-trifluoromethyl-1,2,3,4-tetrahydroquinoxaline-1-methyl-phosphonate (MPQX; ZK 200775; Turski et al. 1996) *spa/spa* rats were implanted subcutaneously minipumps of type 2ML4 (Alzet; alza, Palo Alto, CA, USA). The pumps were reimplanted subcutaneously after 4 weeks.

Two groups of *spa/spa* rats were treated with MPQX. The first group ($n=8$), with an average beginning age of 32 days, received

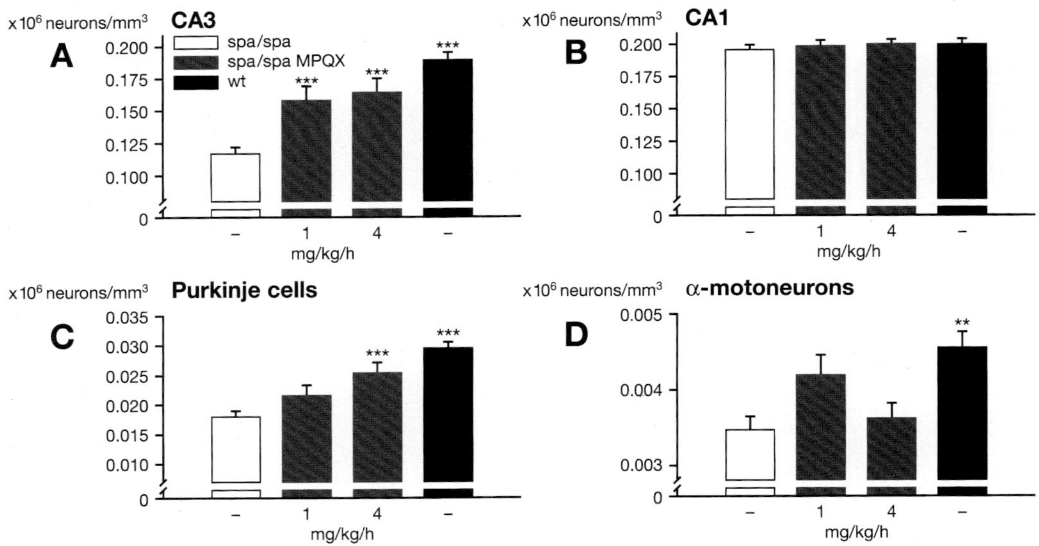

Fig. 6A–D. Neuropathology in genetically determined *spa/spa* rats. An unbiased stereological disector technique was used to calculate the mean numerical density of hippocampal pyramidal neurons, cerebellar Purkinje cells, and spinal α-motoneurons in *spa/spa* (control, n=15; 1 mg kg^{-1} h^{-1} MPQX, n=8; 4 mg kg^{-1} h^{-1} MPQX, n=8) and wildtype rats (n=8). The numerical density for each brain or spinal cord region was determined with 5–10 dissectors per animal. *Bars*, means±SEM of neuronal density. ** $p<0.01$, *** $p<0.001$ vs. *spa/spa* control rats

1 mg kg^{-1} h^{-1}, and the second group (*n*=8), with average beginning age of 31 days, received 4 mg kg^{-1} h^{-1}. For both groups the treatments resulted in higher neuronal density in the hippocampal subfield CA3 than with the vehicle-treated control group, amounting to $0.158\pm0.011\times10^6/mm^3$ (36% increase) and $0.164\pm0.011\times10^6/mm^3$ (40% increase), respectively (Fig. 6A). Treatment with MPQX had no effect on the density of neurons in the hippocampal subfield CA1 (Fig. 6B).

In the cerebellum of *spa/spa* rats which were subjected to chronic treatment with 1 mg kg^{-1} h^{-1} MPQX, the density of Purkinje neurons increased by 20% over that in vehicle-treated *spa/spa* rats, attaining $0.021\pm0.002\times10^6/mm^3$. In rats treated with 4 mg kg^{-1} h^{-1} MPQX Purkinje cell density increased by 40%, attaining $0.025\pm0.002\times10^6/mm^3$ (Fig. 6C).

In the spinal cord the density of α-motoneurons in *spa/spa* rats treated with 1 mg kg^{-1} h^{-1} MPQX attained $0.0042\pm0.0003\times10^6/mm^3$, while the density of α-motoneurons in rats treated with 4 mg kg^{-1} h^{-1} remained at the level of $0.0036\pm0.0002\times10^6/mm^3$ (Fig. 6D).

6.3 Discussion

spa/spa rats carrying a spontaneous autosomal recessive mutation which results in a decrease in the expression of GluR-B mRNA in the brain by about 40% and consequent increase in permeability of AMPA receptors to Ca^{2+}, demonstrate a phenotype which reflects agonist pharmacology of the AMPA receptor. The phenotype in *spa/spa* rats is characterized by age-dependent development of anxiety, increase in muscle tone (spasticity), tremor, progressing motor disturbances, and bulbar palsy leading to death. Neuropathological sequelae in *spa/spa* rats develop over a period of weeks and involve both upper and lower motoneurons. They are most pronounced in spinal α-motoneurons, in the hippocampal CA3 subfield, and in cerebellar Purkinje cells. Long-term treatment with the AMPA antagonist MPQX ameliorates neurodegeneration in the brain and spinal cord of *spa/spa* rats.

The clinical syndrome that we describe in *spa/spa* rats resembles that seen in humans with upper and lower motoneuron disease such as amyotrophic lateral sclerosis. Given the fact that *spa/spa* rats express lower levels of GluR-B mRNA than littermate controls, our data suggest that the resulting high permeability of AMPA receptors to Ca^{2+} causes the chronic neurodegeneration and the clinical symptoms in these animals. This hypothesis is further supported by the demonstration of a neuroprotective effect of a selective AMPA receptor antagonist when administered chronically to these rats.

spa/spa rats developed normally until postnatal day 30, as do the littermates. The phenotype changes starting at the age of about 30 days develop over weeks and reflect the pharmacology of the AMPA receptor (Ikonomidou and Turski 1997). Typically, intracerebroventricular administration of AMPA or systemic administration of α-amino-3-hydroxy-5-tert-butyl-4-isoxazolepropionate (ATPA), a better bioavailable AMPA agonist, to rats or mice causes anxiety, increase in muscle tone and monosynaptic reflexes, and seizures (Turski et al. 1981, 1992). Such changes occur shortly after the administration of AMPA agonists and last for minutes or hours. Administration of ATPA to *spa/spa* rats exacerbated the EMG activity in the gastrocnemius muscle (Turski et al. 1992). Single systemic administration of ATPA did not result in motor disturbances, tremor, or skeletal anomalies such as lordosis or kyphosis in rats. Such changes may be due to long-lasting spasticity and impaired muscle innervation (both consequences of a combination of upper and lower motoneuron degeneration) with resulting loss of muscle mass and development of contractures. Consequences of long-lasting low-dose treatment of rodents with AMPA agonists are not yet known. In such terms, the *spa/spa* rat offers an excellent opportunity to study long-term sequelae of AMPA channel dysfunction which is not severe enough to induce acute changes in behavior. In contrast, transgenic mice expressing unedited GluR-B subunits have a very short survival due to spontaneous seizures and rapid degeneration in the brain (Brusa et al. 1995), which precludes the study of long-term phenotype changes.

6.4 Conclusions

Such experimental evidence suggests that disturbed function of AMPA receptors may be involved in slow-onset neuronal death leading to neurodegenerative disorders. Furthermore, AMPA antagonists which modulate the function of AMPA-sensitive glutamate receptors may delay degenerative processes and offer a therapeutic approach for retarding the progression of neurodegenerative disorders.

References

Brusa R, Zimmermann F, Koh D-S, Feldmeyer D, Gass P, Seeburg P, Sprengel R (1995) Early-onset epilepsy and postnatal lethality associated with an editing-deficient GluR-B allele in mice. Science 270:1677–1680

Clarke KA, Parker AJ (1988) Locomotion in the rat – an underview. Med Sci Res 16:901–902

Cruz-Orive LM, Weibel ER (1990) Recent stereological methods for cell biology: a brief survey. Am J Physiol 258:L148–L156

Gurney ME, Pu H, Chiu AY, Dal Canto MC, Polchow CY, Alexander DD, Caliendo J, Hentati A, Kwon YW, Deng H-X, Chen W, Zhai P, Sufit RL, Siddique T (1994) Motor neuron degeneration in mice that express a human Cu,Zn superoxide dismutase mutation. Science 264:1772–1775

Hawkins LM, Beaver KM, Jane DE, Taylor PM, Sunter DC, Roberts PJ (1995) Characterization of the pharmacology and regional distribution of (S)-^{3}H-5-fluorowillardiine binding in rat brain. Br J Pharmacol 116:2033–2039

Hollmann M, Heinemann S (1994) Cloned glutamate receptors. Annu Rev Neurosci 17:31–108

Ikonomidou C, Turski L (1997) Pharmacology of the AMPA antagonist 2,3-dihydroxy-6-nitro-7-sulfamoyl-(F)-quinoxaline. Ann NY Acad Sci 825:394–402

Klockgether T, Turski L (1990) NMDA antagonists potentiate antiparkinsonian actions of L-dopa in monoamine-depleted rats. Ann Neurol 28:539–546

Margulies JE, Cohen RW, Levine MS, Watson JB (1993) Decreased GluR2(B) receptor subunit mRNA expression in cerebellar neurons at risk for degeneration. Dev Neurosci 15:110–120

Pinna G, Galici R, Schneider HH, Stephens DN, Turski L (1997) Alprazolam dependence prevented by substituting with the β-carboline abecarnil. Proc Natl Acad Sci USA 94:2719–2723

Sommer B, Köhler M, Sprengel R, Seeburg PH (1991) RNA editing in brain controls a determinant of ion flow in glutamate-gated channels. Cell 67:11–19

Turski L, Jacobsen P, Honore T, Stephens D (1992) Relief of experimental spasticity and anxiolytic/anticonvulsant actions of the α-amino-3-hydroxy-5-methyl-4-isoxazolepropionate antagonist 2,3-dihydroxy-6-nitro-7-sulfamoyl-benzo(F)quinoxaline. J Pharmacol Exp Ther 260:742–747

Turski L, Huth A, Sheardown MJ, Jacobsen P, Ottow E (1996) Pharmacology of ZK200775 and ZK202000, competitive non-NMDA glutamate receptor antagonists. Soc Neurosci Abstr 22:1529

Turski WA, Turski L, Czuczwar SJ, Kleinrok Z (1981) (RS)-α-Amino-3-hydroxy-5-methyl-4-isoxazolepropionic acid: wet dog shakes, catalepsy and body temperature changes in rats. Pharmacol Biochem Behav 15:546–549

Wagemann E, Schmidt-Kastner R, Block F, Sontag K-H (1991) Neuronal degeneration in hippocampus and cerebellum of mutant spastic Han-Wistar rats. Neurosci Lett 121:102–106

7 Quantitative Estimation of NMDA Receptor-Mediated Ca^{2+} Entry in Central Neurons

O. Garaschuk, T. Plant, and A. Konnerth

7.1 Introduction

The unique integrative properties of the *N*-methyl-D-aspartate (NMDA) subtype of glutamate receptor channels, including their strong voltage dependence and their high Ca^{2+} permeability, are utilized in a variety of complex physiological mechanisms. So far, NMDA receptor channels have been implicated in the regulation of activity-dependent neuronal development (Constantine-Paton 1990; McDonald and Johnston 1990), in triggering the formation of some forms of synaptic plasticity (Bliss and Collingridge 1993; Malenka and Nicoll 1993), as well as excitatotoxicity (Lipton 1993; Mody and MacDonald 1995) and programmed

A

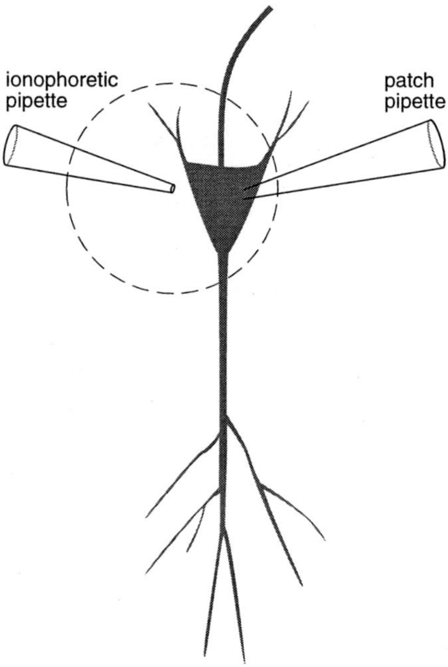

Fig. 1A,B. Measurement of the Ca^{2+} influx through *N*-methyl-D-aspartate (NMDA) receptor channels. **A** Experimental arrangement used in the flux-measurement approach. The patch pipette is filled with the Ca^{2+} indicator dye fura-2 at a concentration of 1–2 mM. The *broken circle* demarcates the area in which the NMDA receptor channels are activated during local ionophoretic NMDA applications. **B** Flux measurements in a hippocampal CA1 pyramidal cell. Traces from *top* to *bottom* are: *a*, fluorescence at 360 nm excitation wavelength (F_{360}); *b*, Ca^{2+}-sensitive fluorescence at 380 nm excitation wavelength (F_{380}); *c*, intracellular free Ca^{2+} concentration ($[Ca^{2+}]_i$), and *d*, whole-cell membrane current (I_{NMDA}). F_{360} is Ca^{2+}-insensitive and was used to estimate the fura-2 concentration inside the cell. The *arrow* indicates the beginning of the whole-cell recording. Here, and in subsequent figures, ionophoretic applications of NMDA were performed at time points marked with *arrowheads*; fluorescence changes are given in bead units (*BU*) representing the fluorescence intensity of a fluorescent bead at 380 nm excitation wavelength (Schneggenburger et al. 1993; Zhou and Neher 1993; Garaschuk and Konnerth 1997). (Modified from Garaschuk et al. 1996). **B** see p. 129

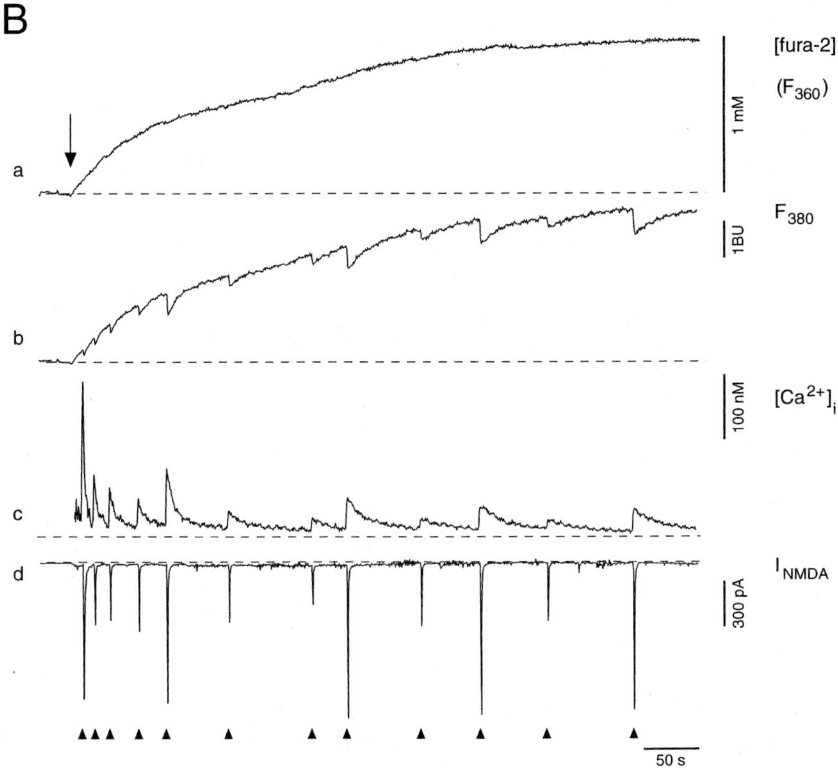

Fig. 1B. Legend see p. 128

cell death (Bonfoco et al. 1995; Qin et al. 1996). Ca^{2+} entry through activated NMDA receptors has been suggested to be involved in the triggering of all of these processes.

Recent molecular biological studies have identified five NMDA receptor (NMDA-R) subunits in rodent brain (NR1 and NR2A–D and ζ1 and ε1–ε4 for the rat and mouse, respectively) and suggested that native NMDA receptor channels are most probably heterooligomers composed of NR1 with one or more of the NR2 subunit types (see Hollmann and Heinemann 1994 for a review). The NR2 subunits are unable to form functional channels alone, but when expressed together with NR1 they increase currents and modify many of the functional properties of the receptors in a subunit-dependent manner. Although the rough mRNA distribution of NMDA receptors in various brain structures is known from in situ hybridization studies (Kutsuwada et al. 1992; Monyer et al. 1992, 1994), the role of different subunits in determining cell-specific functional properties of native NMDA receptors remains largely unknown. This chapter focuses on the comparison of the properties of somatic (extrasynaptic) with those of dendritic (mostly synaptic) NMDA receptors and, in addition, provides an overview of the recent work done to correlate NMDA-R subunit composition and Ca^{2+} permeability of native NMDA receptor channels in central neurons in brain slices.

7.2 Measuring Receptor-Mediated Ca^{2+} Entry Under Physiological Conditions

To measure the NMDA-R-mediated Ca^{2+} entry in neurons in slices we used the fluorometric (fura-2) Ca^{2+} flux-measurement approach initially developed by Neher and Augustine (1992) for studying Ca^{2+} homeostasis in bovine chromaffin cells. This approach makes use of fura-2's property of being not only a Ca^{2+} indicator dye but also a chelator of Ca^{2+} ions. If used at a sufficiently high concentration, fura-2 overrides endogenous Ca^{2+} buffers and reports total Ca^{2+} changes, providing the proper conditions for measurements of Ca^{2+} fluxes. The advantage of the method is that it allows the direct measurement of Ca^{2+} fluxes at physiological extracellular Ca^{2+} concentrations and at negative membrane potentials.

Figure 1A illustrates schematically the experimental arrangement used for Ca^{2+} flux measurements in CA1 pyramidal neurons in hippocampal slices (Garaschuk et al. 1996). To perform fluorometric Ca^{2+} measurements simultaneously with whole-cell patch clamp recordings, patch pipettes were filled with a pipette solution containing 1 mM of the Ca^{2+}-sensitive dye fura-2. NMDA receptor channels were locally activated by ionophoretic application of NMDA from an ionophoresis pipette positioned approximately 5 μm away from the chosen target cell. The *broken circle* in Fig. 1A delineates the approximate area in which the NMDA receptor channels were activated by the agonist. For a given experiment, the size of the circle was estimated by moving the ionophoresis pipette away from the cell and recording electrical and corresponding fluorometric responses as a function of the distance between the cell and the pipette.

As illustrated in Fig. 1B, immediately after establishing the whole-cell recording mode (*arrow*) fura-2 starts to diffuse from the patch pipette into the cell. The fluorescence at the Ca^{2+}-insensitive wavelength F_{360} (Fig. 1Ba) is a direct measure of the intracellular fura-2 concentration and was calibrated accordingly. Shortly after the beginning of the whole-cell recording, NMDA was repeatedly applied to the soma causing brief inward current responses (Fig. 1Bd) and corresponding increases in the free intracellular Ca^{2+} concentration ($[Ca^{2+}]_i$; Fig. 1Bc). Note that as the intracellular fura-2 concentration increased, the amplitude of the change in $[Ca^{2+}]_i$ gradually decreased (Fig. 1Bc), while the amplitudes of the Ca^{2+}-sensitive fluorescence decrements (ΔF_{380}) (Fig. 1Bb) increased. At the end of the loading phase, the exogenous buffer fura-2 largely prevailed over endogenous buffers and ΔF_{380} became dependent exclusively on the amount of NMDA receptor-mediated Ca^{2+} charge entering the cell. Thus, under these experimental conditions single wavelength fluorescence measurements are a quantitative tool for the estimation of Ca^{2+} fluxes.

It is worth noting, however, that the flux-measurement approach critically depends on the quantitative measurements of changes in both the fluorescence and the membrane current. This means that the total fluorescence from a cell under study has to be collected under conditions of stable illumination intensity and constant detection efficiency (for details see Neher 1995) and that for the recording of the membrane current the cell should be under sufficiently good voltage control (Armstrong and Gilly 1992; Müller and Lux 1993), a requirement that is particularly stringent for neurons in slices that preserve their large dendritic trees. In our Ca^{2+} flux-measurement experiments we used the following precautions: (a) we tried to avoid measurements from those sites where there were out-of-focus processes potentially capable of biasing fluorescence measurements; (b) we localized studied Ca^{2+} influx to a well voltage-clamped area of the cell; (c) we blocked all voltage-gated Ca^{2+} channels with broad spectrum Ca^{2+}-channel blocker D-600 (gallopamil); (d) and we blocked or ruled out a contribution of all Ca^{2+}-activated processes (like Ca^{2+}-induced Ca^{2+} release and Ca^{2+}-activated membrane currents).

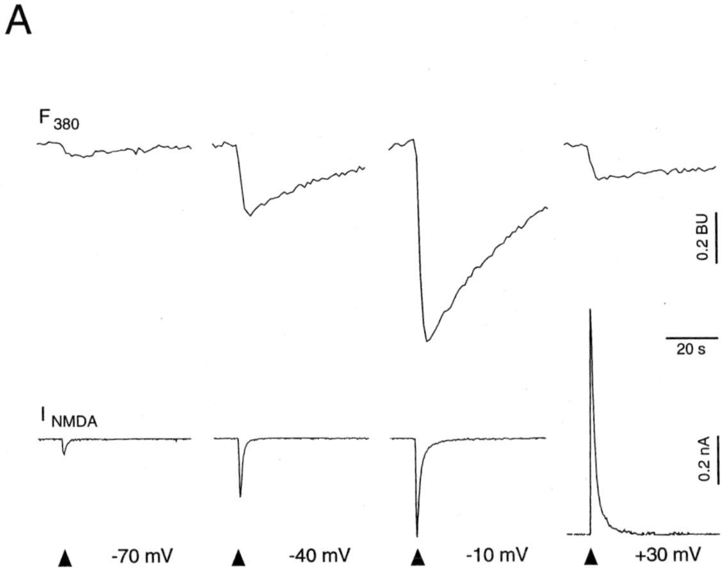

Fig. 2A,B. Voltage-dependence of the Ca^{2+} flux through *N*-methyl-D-aspartate (NMDA) receptor channels. **A** F_{380} traces and corresponding whole-cell currents owing to ionophoretic applications of NMDA (35 ms, 1 μA) at different holding potentials as indicated. Each trace is an average of four consecutive responses. **B** The charge that entered through NMDA receptor channels (Q_{NMDA}), calculated as the time integral of the NMDA-evoked currents, and the corresponding decrements in F_{380} (ΔF_{380}), plotted as a function of holding voltage (V_h). *Error bars* are mean ± SD for four different applications at each holding potential. For ΔF_{380} recordings, the length of the *error bars* is less than the diameter of the symbols. Note that there is significant Ca^{2+} entry even at positive holding voltages (+10–+30 mV) during NMDA receptor-mediated net outward currents. (Taken from Garaschuk et al. 1996). **B** see p. 133

7.3 Ca^{2+} Entry Through Somatic and Dendritic NMDA Receptor Channels in CA1 Pyramidal Neurons

The proportionality between the amount of NMDA receptor-mediated electric charge and the amplitude of the corresponding F_{380} change is illustrated in Fig. 2. Figure 2A shows NMDA-mediated whole-cell currents and the F_{380} responses recorded simultaneously at four different

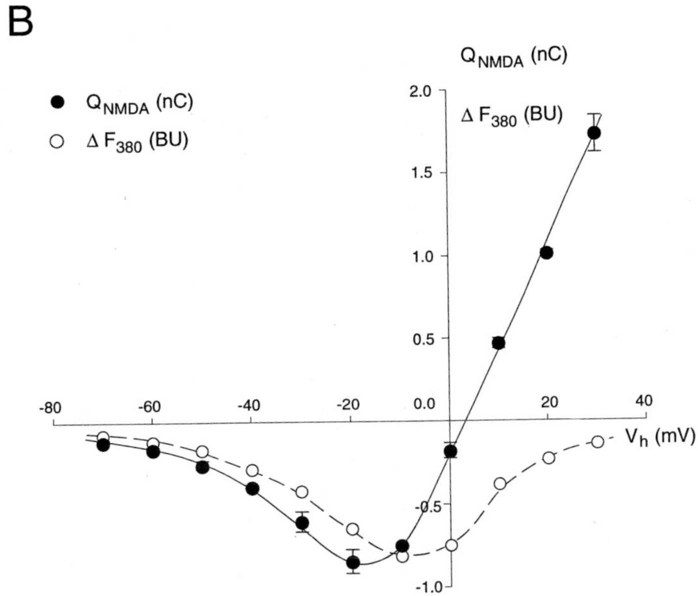

Fig. 2B. Legend see p. 132

holding potentials in standard perfusion saline (1 mM Mg^{2+}). In Fig. 2B the time integrals of the current responses (Q_{NMDA}), representing the net charge crossing the membrane, and corresponding ΔF_{380} values are plotted as a function of membrane potential. Q_{NMDA} responses displayed the well known voltage dependence reported previously to be due to the block by extracellular Mg^{2+} ions (Nowak et al. 1984; Mayer and Westbrook 1987). The Ca^{2+}-sensitive F_{380} decrements closely paralleled the net charge movements at membrane potentials between –70 and –20 mV. It is notable that a significant Ca^{2+} influx through NMDA-activated receptor channels was preserved even when the net cation charge crossing the membrane was negligible or even outward (voltage range, 0–+30 mV). The maximal Ca^{2+}-dependent change in fluorescence was detected at around –10 mV.

The local drug application technique allows comparison of the properties of NMDA receptor channels situated on the soma and dendrites of the same cell. In the experiment illustrated in Fig. 3, after applying

A

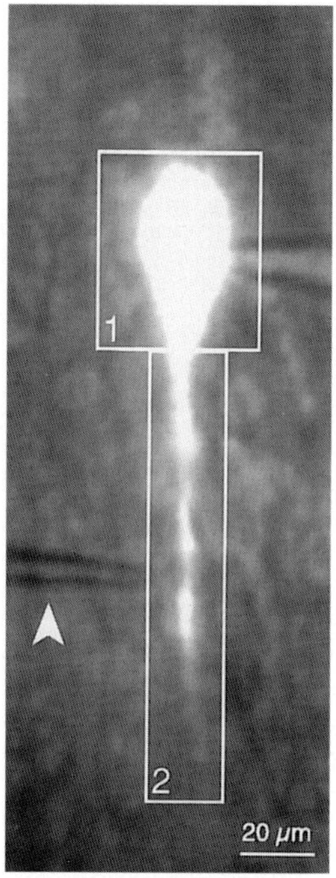

Fig. 3A–C. Localized Ca^{2+} fluxes through somatic and dendritic *N*-methyl-D-aspartate (NMDA) receptor channels of the same cell. **A** A combined epifluorescence and transmitted-light image of a CA1 pyramidal neuron in a hippocampal slice with a somatic patch pipette and a dendritic ionophoresis pipette (*arrowhead*). The fluorescence signals shown in the *top* and *middle traces* of (**B**) and (**C**) were obtained from the somatic (*1*) and the dendritic (*2*) regions, respectively. **B, C** Simultaneously recorded fluorescence (F_{380}) and whole cell current (I_{NMDA}) signals in response to somatic (80–100 ms, 1 μA) and dendritic (30 ms, 1 μA) NMDA applications (*triangles*), respectively. (Modified from Garaschuk et al. 1996). **B,C** see p. 135

NMDA to the soma (Fig. 3B), the ionophoresis pipette was moved to a dendritic region (Fig. 3A, C). In both cases the electrical responses were associated with local F_{380} responses confined specifically to the site of NMDA application (Fig. 3A, B). In general, smaller amounts of NMDA

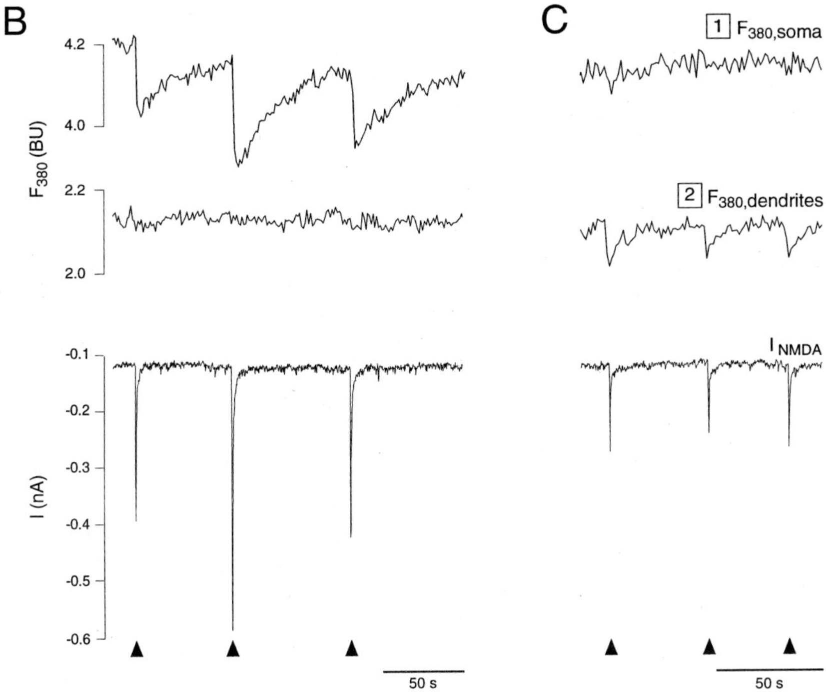

Fig. 3B,C. Legend see p. 134

were applied to the dendrites yielding smaller responses (Fig. 3C) to minimize errors resulting from activation of NMDA receptors located on dendrites that were out of focus. Taking this precaution, consistent results were obtained for all dendritic recordings.

To calculate the fractional Ca^{2+} current, also called P_f, which gives the percentage of total charge contributed by calcium ions, two approaches, illustrated in Fig. 4A, can be used (for detailed description see Schneggenburger et al. 1993; Neher 1995; Garaschuk and Konnerth 1997). In the simplest case, the fraction f that is proportional to P_f can be obtained relating the decrement in fluorescence (ΔF_{380}) at a defined time point to the total ionic charge which entered the cell up to this time (Schneggenburger et al. 1993; Neher 1995).

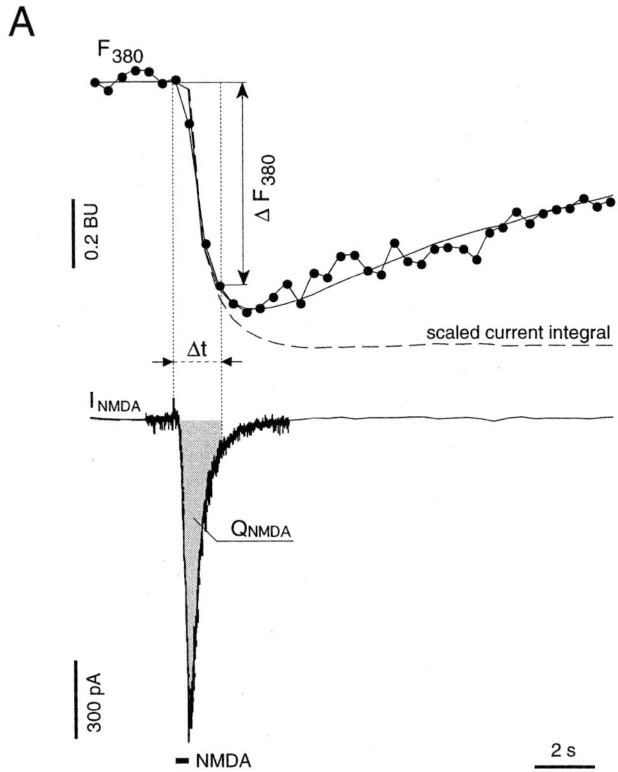

Fig. 4A,B. Comparison of fractional Ca^{2+} currents through somatic and dendritic *N*-methyl-D-aspartate (NMDA) receptor channels in CA1 pyramidal neurons. **A** An NMDA-evoked whole-cell current (*bottom*) and the simultaneously recorded F_{380} trace (*top*). The *broken line* is the scaled time integral of the whole-cell current, the *solid line* is the fluorescence simulation trace (for details see Schneggenburger et al. 1993; Neher 1995). The two approaches available for the calculation of the fraction *f* are illustrated. First, *f* can be obtained by dividing the decrement in fluorescence (ΔF_{380}) by the Q_{NMDA} (*shaded area*). An interval ($\Delta\tau$) delineates a time window during which the F_{380} signal closely follows the scaled integral of the whole-cell current (the *raw data trace* and the *dashed line* coincide) and, therefore, the contribution of the Ca^{2+} extrusion processes to the total fluorescence signal is negligible. Note that just a moment later the *raw data trace* and the *dashed line* start to deviate substantially. Consequently, for later times the $\Delta F_{380}/Q$ approach is no longer applicable. Therefore, the fraction *f* should be obtained from the simulation analysis. **B** see p. 137

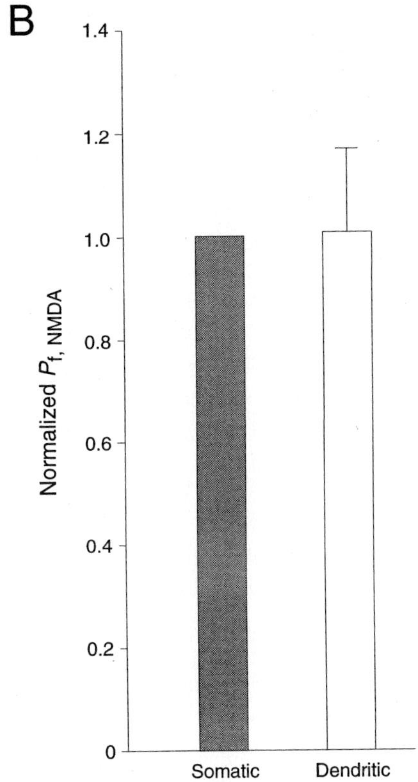

Fig. 4B. Comparison the values of fractional Ca^{2+} current ($P_{f\text{-}NMDA}$) obtained pairwise in somata and dendrites of seven neurons. Each dendritic value was normalized to the corresponding somatic value yielding a mean of 1.01 ± 1.16. (Modified from Garaschuk et al. 1996)

Figure 4A illustrates, however, that the scaled integral of the membrane current, which reflects the Ca^{2+} entry process, starts to deviate substantially from the measured fluorescence signal relatively early (1.7 s) after the initiation of the current. This suggests that even shortly after the onset of Ca^{2+} entry, Ca^{2+} removal processes (including both extrusion and sequestration) contribute significantly to the waveform of the fluorescence signal. Therefore, the contribution of these Ca^{2+} removal processes should be explicitly taken into account. This can be done if the measured fluorescence signal is compared with the simulation of this signal, which assumes, for simplicity, a single time constant Ca^{2+} removal process and is based on the measurement of the corresponding membrane current (Fig. 4A; see also Schneggenburger et al. 1993; Neher 1995). This simulation yields a correct measurement of the fraction f and the time constant of the relaxation process τ. The fractional Ca^{2+} current is then obtained by dividing the measured f value by the coefficient called f_{max}, which gives ΔF_{380} per unit of pure Ca^{2+} charge (Schneggenburger et al. 1993; Garaschuk et al. 1996).

Using the procedures described above to analyze data obtained in CA1 hippocampal pyramidal neurons, we obtained a mean fractional Ca^{2+} current of 10.69% ± 2.13% for somatic NMDA receptor channels at a membrane potential of –60 mV and an extracellular Ca^{2+} concentration of 1.6 mM. As illustrated in Fig. 4B, a very similar value was obtained for the mean fractional Ca^{2+} current through dendritic NMDA receptor channels (10.70% ± 1.96%). In support of these results, similar Ca^{2+} permeabilities, measured from changes in reversal potential, were also found for dendritic and somatic NMDA receptors in membrane patches from CA1 and CA3 pyramidal neurons (Spruston et al. 1995). The Mg^{2+} sensitivity and single-channel conductance of dendritic NMDA receptor channels in CA3 and CA1 pyramidal cells also closely resembled those of the corresponding somatic receptors (Gibb and Colquhoun 1992; Spruston et al. 1995). Moreover, a recent study in cerebellar granule cells (Clark et al. 1997) found no difference between the single-channel properties of somatic and synaptic NMDA receptor channels, strongly suggesting that central neurons do, indeed, express similar receptor channels in their somata and dendritic spines. Interestingly, however, Ito et al. (1997) have recently suggested the existence of synapse-specific targeting of different NMDA receptor subunits (Ito et al. 1997). According to their data, targeted disruption of the gene for NR2A receptor subunit in CA3 hippocampal pyramidal cells of mice selectively reduced NMDA-mediated excitatory postsynaptic currents (EPSCs) and long-term potentiation (LTP) in comissural/associational to CA3 cell synapse, whereas disruption of the gene for NR2B receptor subunit selectively affected NMDA-mediated EPSCs and LTP in fimbrial to CA3 cell synapse (Ito et al. 1997). To our knowledge, this is the first demonstration of the synapse-specific distribution of the NMDA receptor subunits within a single neuron, which may or may not represent a general feature of NMDA receptor distribution in central neurons.

7.4 Ca^{2+} Entry Through NMDA Receptor Channels in GABAergic Neurons of Rat Forebrain

To assess the properties of the NMDA receptor channels in GABAergic (γ-aminobutyric acid) cells, we applied the Ca^{2+} flux-measurement approach to the large GABAergic neurons in the medial septal region (Plant et al. 1997). As illustrated in Fig. 5A, these cells had relatively large somata and thin processes leaving the soma in different directions. The quantitative fluorometric Ca^{2+} measurements in cells of this type may be significantly distorted by the loss of the fluorescence signal coming from thin dendrites that are out-of-focus. Therefore, when analyzing data recorded from these cells, we routinely plotted the estimated P_f values against corresponding values of NMDA receptor-mediated electric charge (Fig. 5B). The data were only taken for further analysis if there was a linear correlation between these two values (a similar analysis was also conducted in CA1 pyramidal cells; Garaschuk et al. 1996). After taking these precautions, the estimated mean fractional Ca^{2+} current for the NMDA receptor channels in medial septal neurons was 12.0% ± 3.9% (Plant et al. 1997).

To ascertain that the analyzed cells represent the population of GABAergic neurons, we used the single-cell reverse transcriptase polymerase chain reaction (RT-PCR) approach (Lambolez et al. 1992). After the fluorescence measurements had been completed, the content of the cell was harvested via the patch pipette (see Plant et al. 1997) and analyzed for the presence of glutamate decarboxylase (GAD).

In conclusion, the established fractional Ca^{2+} current through somatic NMDA receptor channels in GABAergic medial septal neurons was rather similar to that found for glutamatergic CA1 pyramidal neurons (Rogers and Dani 1995; Garaschuk et al. 1996). Interestingly, the P_f values for NMDA receptors in both CA1 pyramidal cells and medial septal neurons were very similar to the P_f values of recombinant NMDA receptors composed of NR1-NR2A subunits expressed in human embryonic kidney (HEK) 293 cells (Burnashev et al. 1995; but see Schneggenburger 1996).

A

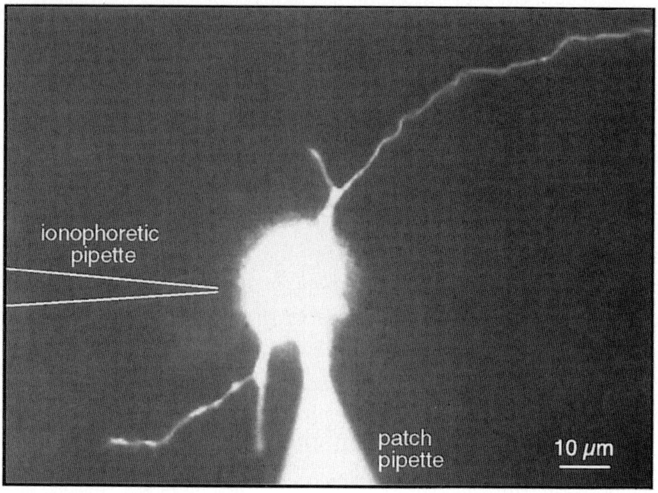

B

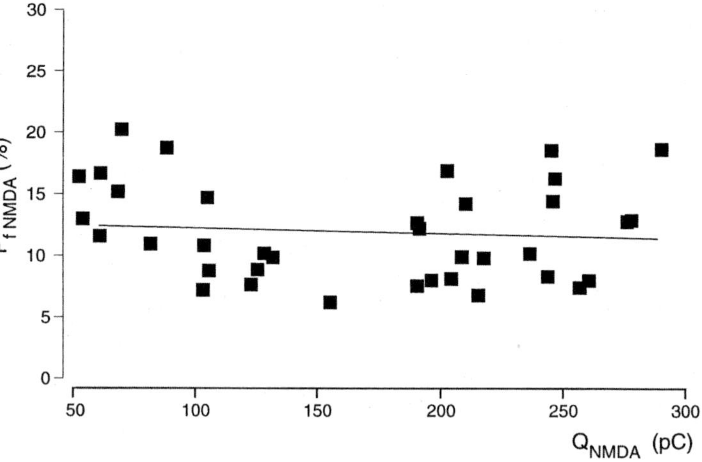

Fig. 5A,B. Legend see p. 141

7.5 Molecular Determinants of the NMDA Receptor Function

To establish possible molecular correlates of the functional properties of native NMDA receptor channels in central neurons we performed single cell RT-PCR analyses of NMDA receptor mRNAs expressed in individual hippocampal and septal neurons (Fig. 6; for details see Garaschuk et al. 1996; Plant et al. 1997). Molecular biological studies have demonstrated that the NR1 subunit is necessary for the formation of functional NMDA receptors and is abundantly expressed throughout all brain structures (see, for example, Monyer et al. 1994; Zukin and Bennett 1995), whereas many of the distinct properties of various NMDA receptors are determined by the NR2 subunits (for review see McBain and Mayer 1994). Therefore, in our experiments we studied the expression pattern of different NR2 subunits and have not analyzed expression of mRNA encoding NR1. In the majority of experiments, we used a set of primers specific for the cDNAs of the NR2A, NR2B, NR2C receptor subunits (see "Methods" in Plant et al. 1997). For technical reasons, we tested for the expression of mRNA for the NR2D subunit in separate series of experiments.

The data shown in Fig. 6A illustrate an experiment in which mRNAs for NR2A and NR2B were detected in a GAD-positive medial septal cell. The results of all experiments in which medial septal cells were investigated for the expression of NR2A-C are summarized in Fig. 6B. NR2B was detected in all but one neuron (i.e. in 96% of neurons), either alone (5/25), together with NR2A (10/25) or NR2C (7/25), or with both NR2A and NR2C (2/25). In the remaining cell, only NR2A was detected. Another group of septal neurons was tested for the expression of NR2D. This subunit was detected in nine out of 12 medial septal neurons (Fig. 6A).

◀ **Fig. 5A,B.** Estimation of the fractional Ca^{2+} current through *N*-methyl-D-aspartate receptors (NMDA-R) in GABAergic (γ-aminobutyric acid) medial septal neurons. **A** An epifluorescence image of a glutamate decarboxylase (GAD)-positive medial septal neuron. **B** Relationship between the values of Pf_{NMDA} and the corresponding ionic charge through NMDA-R channels (Q_{NMDA}) for all cells analyzed in this study. (Modified from Plant et al. 1997)

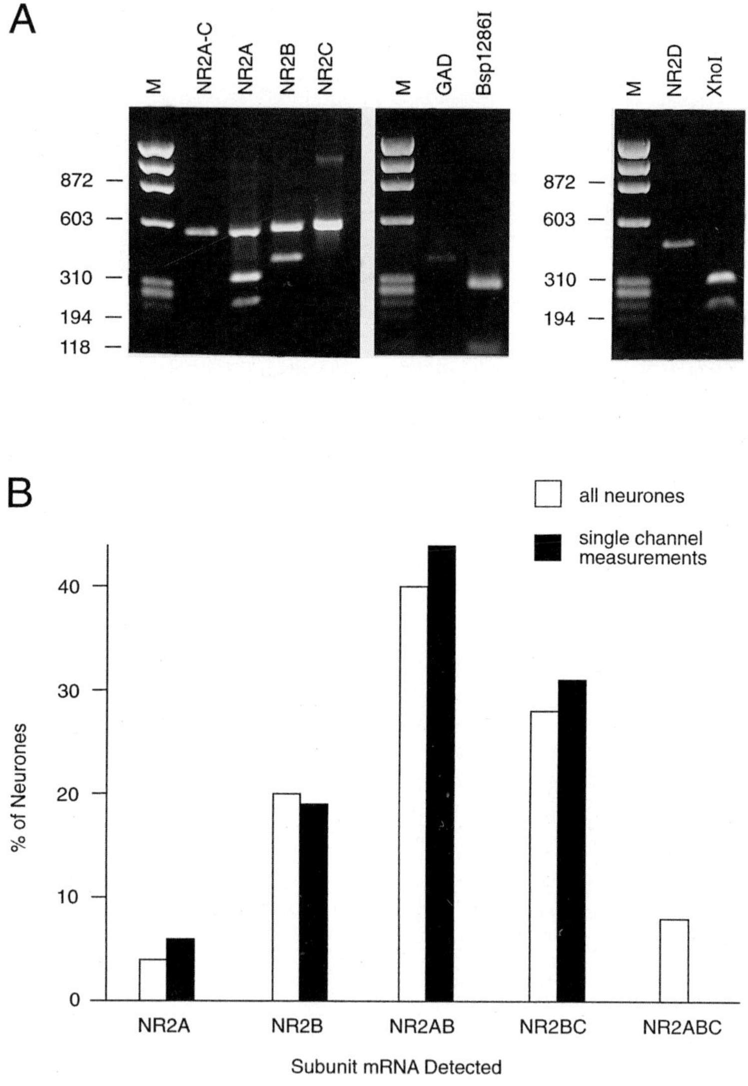

Fig. 6A,B. Legend see p. 143

The NR2B receptor subunit was also predominant in CA1 pyramidal cells (Garaschuk et al. 1996). Altogether we have analyzed 23 pyramidal cells for presence of NMDA receptor-specific mRNAs. NR2A and NR2B specific mRNAs were found in the majority of cells tested. All cells expressed NR2B, two lacked NR2A and two contained NR2A-, NR2B- and NR2C-encoding mRNAs.

Based on these molecular biological studies, it is tempting to conclude that the functional properties of the studied NMDA receptor channels, including their Ca^{2+} permeability, are imparted by the dominant NR2B receptor subunit. Indeed, when this hypothesis was tested in more detail in medial septal neurons, we found that NMDA receptors in these neurons also had a high sensitivity to Mg^{2+} and the NR2B subunit-specific blocker ifenprodil (Plant et al. 1997). Moreover, the main single channel conductance levels observed in patches from GABAergic medial septal neurons were 42 pS and 49 pS in two and 1 mM Ca^{2+}, respectively (Plant et al. 1997). These values were similar to those obtained in recombinant systems when coexpressing NR1 with NR2A or NR2B (50–51 pS; Stern et al. 1992; Tsuzuki et al. 1994). Altogether in our experiments in 2 mM Ca^{2+}, we

◀ **Fig. 6A,B.** Expression of *N*-methyl-D-aspartate receptor (NMDA-R) subunits in medial septal neurons. **A** Agarose gel electrophoresis of cDNA amplified products from medial septal neurons. *Left*, *lane 1*, *M*: Low molecular weight marker *Hae*III digest of ϕX174. *Lane 2*, *single band*: 547-bp amplified product of the first polymerase chain reaction (PCR) for NR2A–NR2C from a single cell. *Lanes 3–5*: Results of digestion with *Bpm*I, *Bfa*I and *Sca*I, specific for NR2A, NR2B and NR2C, respectively, following purification of the product of the first PCR and a second PCR amplification with the same primers. Note that in this neuron NR2A and NR2B were detected. *Centre*, *lane 1*: Marker as above. *Lane 2*: 391-bp product of the PCR amplification with primers for a fragment of glutamate decarboxylase (GAD). *Lane 3*: Restriction digest with *Bsp*1286I after a second PCR amplification. The expected fragments of 291 bp and 100 bp are visible on the gel. *Right*, *lane 1*: As above. *Lane 2*: 465-bp product of the first PCR for NR2D. *Lane 3*: Restriction digest with NR2D specific enzyme *Xho*I. Data for NR2A-C and GAD from the same cell. NR2D from a different cell. **B** Summary of the single cell reverse transcriptase (RT)-PCR data on NR2A-C subunit mRNA expression in medial septal neurons. The patterns of NR2A-C expression in septal neurons are shown for all cells studied (*open columns*, n=25) and for cells in which single channel properties were investigated prior to RT-PCR (*filled columns*, n=16). (Taken from Plant et al. 1997)

Table 1. Comparison of the properties of NMDA-Rs in different neuronal types in brain slices

Type of neuron	Subunit composition	Main single channel conductance (pS)[a] Soma	Dendrites	P_f (%)	References
CA1 pyramidal cell	NR2A, NR2B	51 (44)	– (45)	11	2,3,4,5
GABAergic medial septal neuron	NR2B, NR2D, NR2A, NR2C	49 (42)	– (–)	12	1
Dentate gyrus granule cell	NR2A, NR2B	55 (52)	– (–)	–	2,5,6
Cerebellar granule cell	NR2A, NR2C	52 (–)	58[b] (–)	–	2,7,8,9

1, Plant et al. 1997; 2, Monyer et al. 1994; 3, Spruston et al. 1995; 4, Graraschuk et al. 1996; 5, Strecker et al. 1994; 6, Keller et al. 1991; 7, Ebralidze et al. 1996; 7, Farrant et al. 1994; 8, Ckark et al. 1997.
[a]In outside-out patch.
[b]In cell-attached patch.

observed conductance levels of 51, 43, 31, 19 and 13 pS (Plant et al. 1997), similar values to those obtained under similar conditions in dissociated rat hippocampal neurons (Gibb and Colquhoun 1992). The studies on recombinant receptors (Stern et al. 1992; Tsuzuki et al. 1994; Wyllie et al. 1996) and recent studies in neurons (Farrant et al. 1994; Ebralidze et al. 1996; Momiyama et al. 1996; Takahashi et al. 1996) indicate that the low conductance channels, which in our experiments would correspond to the 19 and 13 pS levels, probably represent conductance levels of NR2C- and NR2D-containing receptors. Thus, conductance levels consistent with the presence of all found NR2 subunits were observed in medial septal neurons. Importantly, however, the 43 pS level, which is consistent with receptors containing NR2B, had the highest probability of opening in all patches, suggesting that it will contribute most to the agonist-induced currents and NMDA-R-mediated component of the EPSC (Plant et al. 1997).

Table 1 provides a comparison of subunit composition and physiological properties of different central neurons studied so far. Because the single-channel conductance of NMDA receptor channels depends on the external Ca^{2+} concentration (Ascher and Nowak 1988; Gibb and Colquhoun 1992; Jahr and Stevens 1993), two values reflecting main single channel conductance in 1- and 2-mM external Ca^{2+} are given.

The value of 58 pS found in cerebellar granule cells represents the main single channel conductance of synaptic receptor channels recorded in cell-attached patches (Clark et al. 1997). This value is very similar to the one found for somatic NMDA receptor channels under similar experimental conditions (64 pS; Clark et al. 1997). The P_f value of 11% found in CA1 pyramidal cells applies to both somatic and dendritic NMDA receptor channels (Garaschuk et al. 1996).

Comparison of the functional properties of the listed somatic and dendritic NMDA receptor channels with the properties of recombinant NMDA receptor channels (Monyer et al. 1992, 1994; Stern et al. 1992; Tsuzuki et al. 1994; Wyllie et al. 1996) allows the conclusion that, in spite of differences in the NR2 subunit combination, most of these properties in the best studied types of central neurons, including glutamatergic hippocampal CA1 pyramidal neurons and GABAergic medial septal neurons, are similar and correlate with the abundant expression of NR2B and NR2A receptor subunits.

7.6 Conclusions

In conclusion, the combination of whole-cell patch clamp recordings, Ca^{2+} flux measurements and RT-PCR analyses on the level of single identified neurons in brain slices is a powerful tool for understanding the role of subunit composition in determining functional properties of NMDA receptor channels. We found that in hippocampal CA1 pyramidal neurons the properties of dendritic, presumably synaptic, NMDA receptor channels resemble those of somatic extrasynaptic NMDA receptor channels and are consistent with the abundant expression of the NR2A/NR2B receptor subunits. In the GABAergic septal neurons, however, the cellular function seems to be determined by the expression of the NR2B subunit. Finally, the Ca^{2+} flux-measurement approach may be used also to study the Ca^{2+} entry process through NMDA receptor channels with a high spatial resolution, including that through dendritic spines, and thus to quantify the Ca^{2+} entry necessary for triggering various NMDA receptor-mediated physiological mechanisms.

Acknowledgements. This work was supported by grants from the DFG (SFB 246), the BMFB and INTAS (94–4072).

References

Armstrong CM, Gilly WF (1992) Access resistance and space clamp problems associated with whole-cell patch clamping. Methods Enzymol 207:100–122

Ascher P, Nowak L (1988) The role of divalent cations in the N-methyl-D-aspartate responses of mouse central neurones in culture. J Physiol (Lond) 399:247–266

Bliss TVP, Collingridge GL (1993) A synaptic model of memory: long-term potentiation in the hippocampus. Nature 361:31–39

Bonfoco E, Krainc D, Ankarcrona M, Nicotera P, Lipton SA (1995) Apoptosis and necrosis: two distinct events induced, respectively, by mild and intense insults with N-methyl-D-aspartate or nitric oxide/superoxide in cortical cell cultures. Proc Natl Acad Sci USA 92:7162–7166

Burnashev N, Zhou Z, Neher E, Sakmann B (1995) Fractional calcium currents through recombinant GluR channels of the NMDA, AMPA and kainate receptor subtypes. J Physiol (Lond) 485:403–418

Clark BA, Farrant M, Cull Candy SG (1997) A direct comparison of the single-channel properties of synaptic and extrasynaptic NMDA receptors. J Neurosci 17:107–116

Constantine-Paton M (1990) NMDA receptor as a mediator of activity-dependent synaptogenesis in the developing brain. Cold Spring Harb Symp Quant Biol 55:431–443

Ebralidze AK, Rossi DJ, Tonegawa S, Slater NT (1996) Modification of NMDA receptor channels and synaptic transmission by targeted disruption of the NR2C gene. J Neurosci 16:5014–5025

Farrant M, Feldmeyer D, Takahashi T, Cull Candy SG (1994) NMDA-receptor channel diversity in the developing cerebellum. Nature 368:335–339

Garaschuk O, Konnerth A (1997) Quantitative calcium imaging in brain slices. In: Rizzuto R, Fasolato C (eds) Imaging living cells. Landes/Springer, Austin (in press)

Garaschuk O, Schneggenburger R, Schirra C, Tempia F, Konnerth A (1996) Fractional Ca^{2+} currents through somatic and dendritic glutamate receptor channels of rat hippocampal CA1 pyramidal neurones. J Physiol (Lond) 491:757–772

Gibb AJ, Colquhoun D (1992) Activation of N-methyl-D-aspartate receptors by L-glutamate in cells dissociated from adult rat hippocampus. J Physiol (Lond) 456:143–179

Hollmann M, Heinemann S (1994) Cloned glutamate receptors. Annu Rev Neurosci 17:31–108

Ito I, Futai K, Katagiri H, Watanabe M, Sakimura K, Mishina M, Sugiyama H (1997) Synapse-selective impairment of NMDA receptor functions in mice lacking NMDA receptor ε1 or ε2 subunit. J Physiol (Lond) 500 (2):401–408

Jahr GE, Stevens CF (1993) Calcium permeability of the N-methyl-D-aspartate receptor channel in hippocampal neurons in culture. Proc Natl Acad Sci USA 90:11573–11577

Keller BU, Konnerth A, Yaari Y (1991) Patch clamp analysis of excitatory synaptic currents in granule cells of rat hippocampus. J Physiol (Lond) 435:275–293

Kutsuwada T, Kashiwabuki N, Mori H, Sakimura K, Kushiya E, Araki K, Meguro H, Masaki H, Kumanishi T, Arakawa M, Mishina M (1992) Molecular diversity of the NMDA receptor channel. Nature 358:36–41

Lambolez B, Audinat E, Bochet P, Crepel F, Rossier J (1992) AMPA receptor subunits expressed by single Purkinje cells. Neuron 9:247–258

Lipton (1993) Prospects for clinically tolerated NMDA antagonists: open channel blockers and alternative redox states of nitric oxide. Trends Neurosci 16:527–532

Malenka R, Nicoll R (1993) NMDA-receptor-dependent synaptic plasticity: multiple forms and mechanisms. Trends Neurosci 16:521–526

Mayer ML, Westbrook GL (1987) The physiology of excitatory amino acids in the vertebrate central nervous system. Prog Neurobiol 28:197–276

McBain CJ, Mayer ML (1994) N-methyl-D-aspartic acid receptor structure and function. Physiol Rev 74:723–760

McDonald JW, Johnston MV (1990) Physiological and pathophysiological roles of excitatory amino acids during central nervous system development. Brain Res Brain Res Rev 15:41–70

Mody I, MacDonald JF (1995) NMDA receptor-dependent excitotoxicity: the role of intracellular Ca^{2+} release. Trends Pharmacol Sci 16:356–359

Momiyama A, Feldmeyer D, Cull Candy SG (1996) Identification of a native low-conductance NMDA channel with reduced sensitivity to Mg^{2+} in rat central neurones. J Physiol (Lond) 494:479–492

Monyer H, Sprengel R, Schoepfer R, Herb A, Higuchi M, Lomeli H, Burnashev N, Sakmann B, Seeburg PH (1992) Heteromeric NMDA receptors: molecular and functional distinction of subtypes. Science 256:1217–1221

Monyer H, Burnashev N, Laurie DJ, Sakmann B, Seeburg PH (1994) Developmental and regional expression in the rat brain and functional properties of four NMDA receptors. Neuron 12:529–540

Müller W, Lux HD (1993) Analysis of voltage-dependent membrane currents in spatially extended neurons from point-clamp data. J Neurophysiol 69:241–247

Neher E (1995) The use of fura-2 for estimating Ca buffers and Ca fluxes. Neuropharmacology 34:1423–1442

Neher E, Augustine GJ (1992) Calcium gradients and buffers in bovine chromaffin cells. J Physiol (Lond) 450:273–301

Nowak L, Bregestovski P, Ascher P, Herbet A, Prochiantz A (1984) Magnesium gates glutamate-activated channels in mouse central neurones. Nature 307:462–465

Plant T, Schirra C, Garaschuk O, Rossier J, Konnerth A (1997) Molecular determinants of NMDA receptor function in GABAergic neurones of rat forebrain. J Physiol (Lond) 499:47–63

Qin ZH, Wang Y, Chase TN (1996) Stimulation of N-methyl-D-aspartate receptors induces apoptosis in rat brain. Brain Res 725:166–176

Rogers M, Dani JA (1995) Comparison of quantitative calcium flux through NMDA, ATP, ACh receptor channels. Biophys J 68:501–506

Schneggenburger R (1996) Simultaneous measurement of Ca^{2+} influx and reversal potentials in recombinant N-methyl-D-aspartate receptor channels. Biophys J 70:2165–2174

Schneggenburger R, Zhou Z, Konnerth A, Neher E (1993) Fractional contribution of calcium to the cation current through glutamate receptor channels. Neuron 11:133–143

Spruston N, Jonas P, Sakmann B (1995) Dendritic glutamate receptor channels in rat hippocampal CA3 and CA1 pyramidal neurons. J Physiol (Lond) 482:325–352

Stern P, Behe P, Schoepfer R, Colquhoun D (1992) Single-channel conductances of NMDA receptors expressed from cloned cDNAs: comparison with native receptors. Proc R Soc Lond B Biol Sci 250:271–277

Strecker GJ, Jackson MB, Dudek FE (1994) Blockade of NMDA-activated channels by magnesium in the immature rat hippocampus. J Neurophysiol 72:1538–1548

Takahashi T, Feldmeyer D, Suzuki N, Onodera K, Cull Candy SG, Sakimura K, Mishina M (1996) Functional correlation of NMDA receptor epsilon subunits expression with the properties of single-channel and synaptic currents in the developing cerebellum. J Neurosci 16:4376–4382

Tsuzuki K, Mochizuki S, Iino M, Mori H, Mishina M, Ozawa S (1994) Ion permeation properties of the cloned mouse epsilon 2/zeta 1 NMDA receptor channel. Brain Res Mol Brain Res 26:37–46

Wyllie DJ, Behe P, Nassar M, Schoepfer R, Colquhoun D (1996) Single-channel currents from recombinant NMDA NR1a/NR2D receptors expressed in Xenopus oocytes. Proc R Soc Lond B Biol Sci 263:1079–1086

Zhou Z, Neher E (1993) Calcium permeability of nicotinic acetylcholine receptor channels in bovine adrenal chromaffin cells. Pflugers Arch 425:511–517

Zukin RS, Bennett MV (1995) Alternatively spliced isoforms of the NMDARI receptor subunit. Trends Neurosci 18:306–313

8 Calcium Permeability of Glutamate-Gated Channels in the CNS: Functional Role of the Ca^{2+}-Permeable AMPA Receptor Channels

N. Burnashev

8.1 Homologous Q/R/N-Site in M2 Region Controls Ca^{2+} Permeability in GluR Channels

Ionotropic glutamate receptor (GluR) channels are assembled from structurally homologous subunits that can be grouped into three families according to sequence identity and agonist binding characteristics. These are *N*-methyl-D-aspartate receptor (NMDAR), α-amino-3-hydroxy-5-methyl-4-isoxazolepropionate receptor (AMPAR), and kainate

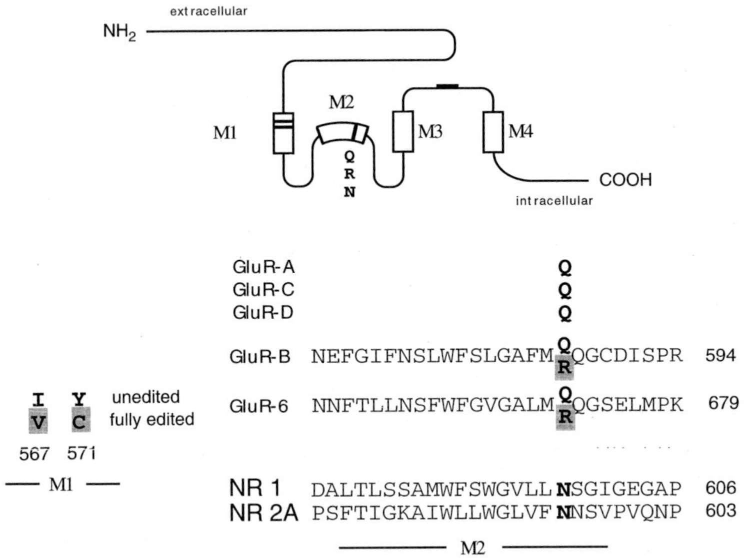

Fig. 1. Linear representation of the glutamate receptor (GluR) sequence. Sequences for the pore-forming loop, M2, are shown in detail. The homologous positions in M2 corresponding to the Q/R/N-site and editing sites in the M1 region of the GluR6 subunit are highlighted

receptor (KAR) channels. The latter two correspond to traditionally classified non-NMDA receptors. The primary structure of the cloned subunits revealed prominent structural similarities between all receptor types. The subunits are about 1000 amino acids in length and have four hydrophobic segments, M1–M4 (Hollmann and Heinemann 1994). Experimental evidence for the AMPAR subunits (Hollmann et al. 1994; Bennett and Dingledine 1995) and the NMDAR subunits (Wood et al. 1995; Kuner et al. 1996) strongly suggests that the M2 region of GluR forms a re-entrant membrane loop with both ends facing the cytoplasm (Fig. 1). The exact topology of GluRs is not known, but a growing body of evidence indicates that the amphipathic M2 segment is a part of GluR channel's pore (for review see Hollmann and Heinemann 1994). Functional analysis of recombinantly expressed GluR channels revealed a homologous site in the M2 region – the Q/R/N-site – which controls conductance and ion selectivity of these channels.

In several subunits this site is spot-changed by RNA editing. For one of the AMPAR subunits, GluR-B, the respective genomic exon predicts the presence of a glutamine (Q) residue at the site and arginine (R) is introduced into this site by RNA editing (Sommer et al. 1991). GluR5 and GluR6 subunits of high affinity KAR channels also carry a glutamine or arginine residue in a homologous site of the putative channel forming segment M2. The glutamine is encoded by the gene, and the arginine is again generated by RNA editing (Sommer et al. 1991).

8.1.1 Ca^{2+} Permeability and Polyamine Block of the Q- and R-Form AMPAR Channels

AMPAR channels assembled from glutamine containing subunits, and known as Q-form channels, show high Ca^{2+} permeability ($P_{Ca}/P_{monovalent}$=2) as measured from reversal potentials in high Ca^{2+} extracellular solutions, and relatively large fractional Ca^{2+} current (4%) with physiological Ca^{2+} concentration estimated from the photometric measurements of Ca^{2+} flux. They are also blocked by naturally occurring intracellular polyamines in a voltage-dependent manner, generating a doubly rectifying current–voltage relation (I-V). The doubly rectifying whole-cell I-V that is a hallmark of Ca^{2+}-permeable AMPAR channels, becomes linear in outside-out patches after washout of an intracellular content. Addition into the recording pipette of spermine in micromolar concentrations makes outside-out I-V again doubly rectifying. AMPARs containing R-form subunits show low Ca^{2+} permeability ($P_{Ca}/P_{monovalent}$ =0.1) and fractional Ca^{2+} current (0.5%). They are insensitive to polyamines and have linear I-Vs, both in whole-cell and outside-out recording configurations (for review see Jonas and Burnashev 1995). Low Ca^{2+} permeability and insensitivity to polyamine blockade of the R-form channels is consistent with the notion that positively charged arginine provides a barrier for divalent cations to permeate through the channel and for positively charged polyamines to reach their blocking site due to electrostatic interaction.

8.1.2 Kainate Receptor Channels

The Q- and R-forms of GluR6 channels show similar sensitivity to polyamine blockade. The whole-cell I-V of Q-form channels is also doubly rectifying, whereas the I-V of R-form channels is linear (Bowie and Mayer 1995; Kamboj et al. 1995). However, measurements of the reversal potentials under biionic conditions at high Ca^{2+} concentrations indicated that homomeric R-form channels expressed in human embryonic kidney (HEK) 293 cells were more permeable to Ca^{2+} than Q-form channels (Egebjerg and Heinemann 1993; Köhler et al. 1993). On the other hand, direct measurements of the Ca^{2+} influx using Ca^{2+}-sensitive dye, fura-2, have shown that the fractional Ca^{2+} current through homomeric R-form channels is very small – less than 0.2 % of the total charge transfer. This value is smaller than that for homomeric Q-form channels and heteromeric channels assembled from the Q- and R-form subunits and is the smallest value of all GluR subunits measured (Burnashev et al. 1995).

Experiments with CsCl concentration gradients indicated that the homomeric R-form channels from the GluR6 subunit are about equally permeable for Cs^+ and Cl^- (Burnashev et al. 1996). Thus, the relatively high Cl^- permeability of the homomeric GluR6(R) channels (P_{Cl}/P_{Cs}=0.74) explains a discrepancy in Ca^{2+} permeability estimated from biionic Ca^{2+}/Cs^+ reversal potentials on the one hand (Egebjerg and Heinemann 1993, Köhler et al. 1993), and from direct measurement of Ca^{2+} influx using photometric techniques on the other (Burnashev et al. 1995). In the first case, the apparent reversal potential for Ca^{2+} was occluded by a contribution of Cl^- to the total current, resulting in an overestimation of Ca^{2+} permeability. In the second case, photometric measurements indicated a low level of Ca^{2+} influx reflecting an actual low Ca^{2+} permeability of these channels. Interestingly the P_{Cl}/P_{Cs} value in high Ca^{2+} extracellular solutions increased almost two-fold compared to the value obtained in CsCl gradient experiments. This is consistent with the notion that low permeant Ca^{2+} ions contribute to the total positive charge distribution near the narrow constriction of the channel, lowering a barrier for the entry of anions. Noteworthy is that homomeric edited GluR-B(R) channels also showed detectable Cl^- permeability but the P_{Cl}/P_{Cs} was 0.14 – smaller than that for GluR6(R) channels (Burnashev et al. 1996).

Further analysis of cloned cDNAs revealed that in the GluR6 subunit two additional positions, located in transmembrane segment M1, are diversified by RNA editing to generate either isoleucine (I) or valine (V) in one, and tyrosine (Y) or cysteine (C) in the other M1 position (see Fig. 1). M1 editing also affects Ca^{2+} permeability of GluR6 channels. Thus, GluR6(Q) channels fully edited in the M1 segment showed a reduced Ca^{2+} permeability compared to that of the M1 unedited subunit (Köhler et al. 1993).

8.1.3 NMDA Receptor Channels

The NMDAR subunits NR1 and NR2 (A–D) contain an asparagine (N) at a site in the M2 region (N-site) which is homologous to the editing site in the M2 region of non-NMDAR channels. Wild-type NMDAR channels are highly permeable to Ca^{2+}. Introducing a glutamine or arginine to the N-site of an NR1 subunit reduced the Ca^{2+} permeability, and conductance of the channels composed of NR1 and one of the NR2 subunits (Burnashev et al. 1992c), indicating that this site contributes to the ion permeation of the NMDAR channels.

8.2 Relative Ca^{2+} Permeability of GluR Channel Subtypes in the CNS

Excitatory synaptic transmission in the mammalian CNS is mediated by NMDAR and AMPAR channels. These two subtypes of GluRs contribute differently to the excitatory post-synaptic current (EPSC). Thus, AMPARs mediate the fast component of EPSCs, whereas NMDARs mediate currents which are slower and longer lasting (Bekkers and Stevens 1989; Stern et al. 1992). Comparison of the fractional Ca^{2+} currents through different subtypes of NMDAR channels shows a relatively small variation in these values (8%–12%). In contrast, fractional Ca^{2+} currents through AMPAR channels vary over a wide range (0.5%–4%) (Burnashev et al. 1995). Studies of recombinantly expressed AMPAR channels led to the hypothesis that differential expression levels of the GluR-B subunit in its edited form might generate an assembly of a mosaic of various AMPAR channels with different Ca^{2+}

permeability within individual cells (Burnashev et al. 1992b). This mechanism appears to operate in native neurons. Thus, in cerebellar Bergmann glial cells, which do not express GluR-B subunit, as was shown by in situ hybridization studies on slices and cultured cells, AMPARs displayed doubly rectifying whole-cell I-V and high Ca^{2+} permeability (Burnashev et al. 1992a; Müller et al. 1992). In contrast, cerebellar granule cells containing edited GluR-B subunits showed linear I-V and low Ca^{2+} permeability (Burnashev et al. 1992a).

Several recent studies have shown that in different identified neurons the Ca^{2+} permeability of AMPAR channels varies over a wide range (reviewed in Jonas and Burnashev 1995). Probing different CNS cells using patch-clamp recordings of AMPAR-mediated currents, combined with analysis of mRNA in the same cell using reverse transcription followed by the polymerase chain reaction, revealed a highly significant inverse correlation between the Ca^{2+} permeability of native AMPARs and the relative abundance of GluR-B mRNA (Geiger et al. 1995). Assuming that the relative abundances of GluR-B mRNA and GluR-B subunit proteins correlate closely, any intermediate Ca^{2+} permeability could be generated in a given cell by varying the relative abundance of GluR-B mRNA (Geiger et al. 1995).

8.3 Modulation of the NMDAR Channels by Fast Ca^{2+} Entry Through Co-localized AMPAR Channels

Comparison of the relative contribution of AMPAR channels with high Ca^{2+} permeability to the Ca^{2+} influx during synaptic transmission indicates that at short integration time intervals the fraction of Ca^{2+} ions mediated by AMPAR channels might be comparable or larger than that by NMDARs (Koh et al. 1995b).

It has been shown that an increase in the free intracellular Ca^{2+} concentration ($[Ca^{2+}]_i$) causes a reversible inactivation of NMDAR channels irrespective of the source of Ca^{2+} elevation (Legendre et al. 1993; Viklicky 1993; Medina et al. 1994; Kyrozis et al. 1995). To examine the interaction between AMPARs and NMDARs, Ca^{2+}-permeable GluR-B(Q) AMPAR and Ca^{2+}-impermeable NR1(R)-NR2A NMDAR channels were co-expressed in HEK 293 cells. This meant that self-inhibition of NMDARs by Ca^{2+} influx through NMDAR channels could be excluded. Submicro-molar $[Ca^{2+}]_i$ arising from AMPARs reversibly reduced NMDAR whole-cell currents, as revealed by simultaneous measurements of the whole-cell currents and $[Ca^{2+}]_i$. To mimic the evoked post-synaptic current, a 1-ms pulse of 3 m*M* glutamate was applied to outside-out patches pulled from the cells co-expressing GluR-B(Q) and NR1(R)-NR2A channels. The recorded currents had fast (AMPAR) and slow (NMDAR) components resembling those of EPSCs. At 10 mM extracellular of Ca^{2+} the peak of the NMDA component of the glutamate-evoked current was reduced on average by 25%, compared to that in the presence of CNQX, a selective blocker of AMPARs (Fig. 2A). Thus, Ca^{2+} entering through the Ca^{2+}-permeable AMPARs might inactivate co-localized NMDARs in the time range of excitatory post-synaptic currents (Rozov et al. 1997a).

8.4 Self-Facilitation of Ca^{2+}-Permeable AMPAR Channels by Activity-Dependent Relief from Polyamine Block

In outside-out patches excised from HEK 293 cells expressing the Ca^{2+}-permeable AMPAR subunits GluR-A, GluR-B(Q) or GluR-D, a 1-ms pulse of 3 mM glutamate activated fast decaying currents which facilitated upon repetitive (2–33 Hz) agonist application (Fig. 2B). In contrast, Ca^{2+}-impermeable heteromeric GluR-A/B(R) channels did not facilitate at any stimulation frequency (0.5–100 Hz). Although facilitation was observed only in Ca^{2+}-permeable AMPARs, Ca^{2+} influx per se, however, is not required to induce facilitation, since facilitation in these channels might be observed also in Ca^{2+}-free extracellular solution.

Ca^{2+}-permeable AMPAR channels are blocked by endogenous polyamines in a voltage-dependent manner, producing a characteristic doubly rectifying I-V. In outside-out patches, the I-V of polyamine-sensitive AMPAR channels linearizes over time due to washout of intracellular polyamines (Bowie and Mayer 1995; Kamboj et al. 1995; Koh et al. 1995a). However, high Ca^{2+} permeability and sensitivity to polyamine block may be dissociated by substituting asparagine (N) (Burnashev et al. 1992b) at the Q/R-site in the M2 region of AMPAR subunits. Mutant GluR-B(N) channels have a high Ca^{2+} permeability, but linear I-V, as a result of insensitivity to polyamine block (Burnashev et al. 1992b; Koh et al. 1995a), and do not show facilitation (Fig. 2C).

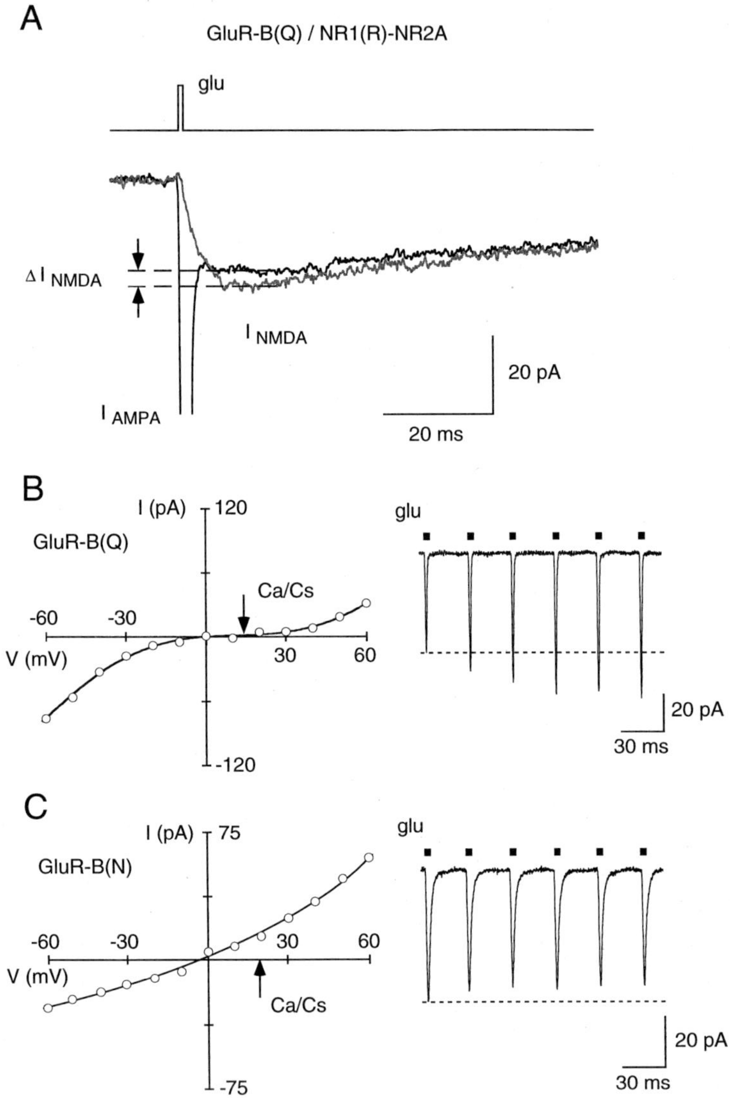

Fig. 2. Legend see p. 157

This suggests that current facilitation does not depend on Ca^{2+} influx but may relate to polyamine sensitivity.

Further experiments showed that this facilitation arises by the use-dependent relief of the block by polyamines (Rozov et al. 1997b). Firstly, after washout of the intracellular content in outside-out configuration, the facilitation disappeared upon linearization of the I-V. In the presence of spermine in the recording pipette, the shape of the I-V and facilitation properties did not change significantly during the recording time. Secondly, the I-V for currents facilitated by a conditioning train of glutamate pulses becomes more linear, suggesting that partial relief of AMPAR channels from polyamine block underlies the enhancement of the currents. Finally, the degree of facilitation showed a pronounced voltage dependence which paralleled the degree of block by polyamines. Thus, polyamine sensitivity of Ca^{2+}-permeable AMPAR channels, but not Ca^{2+} entry, underlies facilitation.

Experiments on patches from rat hippocampal slices revealed that facilitation occurs in dentate gyrus basket cells but not in CA1 pyramidal cells. These cell types express polyamine-sensitive and -insensitive AMPAR subunits, respectively (Koh et al. 1995a). The results suggest that in neurons expressing Ca^{2+}-permeable AMPAR channels

---

◀ **Fig. 2A–C.** Functional role of the Ca^{2+}-permeable α-amino-3-hydroxy-5-methyl-4-isoxazolepropionate receptor (AMPAR) channels. **A** Fast Ca^{2+} entry through the AMPAR channels inhibits the *N*-methyl-D-aspartate receptor (NMDAR) component of the currents activated by a 1-ms pulse of 3 mM glutamate applied to an outside-out patch from human embryonic kidney (HEK) 293 cell co-expressing Ca^{2+}-permeable GluR-B(Q) and Ca^{2+}-impermeable NR1(R)–NR2A channels. Traces were recorded from the same patch in control extracellular solution with 10 mM Ca^{2+} (*black line*) and in the presence of 10 µM CNQX (*gray line*). Membrane potential, –60 mV. ΔI_{NMDA}, the difference in the NMDAR component amplitudes measured with and without CNQX. **B** Polyamine-sensitive AMPAR GluR-B(Q) channels have doubly rectifying I-V (*left*) and show gradual current facilitation during a train of 1-ms glutamate (3 m*M*) pulses applied at 33 Hz. Outside-out patches with spermine added to the recording pipettes were carried out. Membrane potential, –60 mV. Extracellular solution used was normal rat Ringer's solution. **C** Ca^{2+}-permeable polyamine-insensitive mutant AMPAR GluR-B(N) channels have linear I-V (*left*) and do not show current facilitation under the same experimental conditions. *Arrows* indicate the Ca^{2+}/Cs^{+} reversal potentials measured with high Ca^{2+} (110 m*M*) extracellular solutions polyamine-AMPAR interaction may function as an effective post-synaptic mechanism in neuronal plasticity.

References

Bekkers JM, Stevens CF (1989) NMDA and non-NMDA receptors are co-localized at individual excitatory synapses in cultured rat hippocampus. Nature 341:230–233

Bennett JA, Dingledine R (1995) Topology profile for a glutamate receptor: three transmembrane domains and a channel-lining reentrant membrane loop. Neuron 14:373–384

Bowie D, Mayer ML (1995) Inward rectification of both AMPA and kainate subtype glutamate receptors generated by polyamine-mediated ion channel block. Neuron 15:453–462

Burnashev N, Khodorova A, Jonas P, Helm PJ, Wisden W, Monyer H, Seeburg PH, Sakmann B (1992a) Calcium-permeable AMPA-kainate receptors in fusiform cerebellar glial cells. Science 256:1566–1570

Burnashev N, Monyer H, Seeburg PH, Sakmann B (1992b) Divalent ion permeability of AMPA receptor channels is dominated by the edited form of a single subunit. Neuron 8:189–198

Burnashev N, Schoepfer R, Monyer H, Ruppersberg JP, Günther W, Seeburg PH, Sakmann B (1992c) Control by asparagine residues of calcium permeability and magnesium blockade in the NMDA receptor. Science 257:1415–1419

Burnashev N, Zhou Z, Neher E, Sakmann B (1995) Fractional calcium currents through recombinant GluR channels of the NMDA, AMPA and kainate receptor subtypes. J Physiol 485:403–418

Burnashev N, Villarroel A, Sakmann B (1996) Dimensions and ion selectivity of recombinant AMPA and kainate receptor channels and their dependence on Q/R site residues. J Physiol 496:165–173

Egebjerg J, Heinemann, SF (1993) Ca^{2+} permeability of unedited and edited versions of the kainate-selective glutamate receptor GluR6. Proc Natl Acad Sci USA 90:755–759

Geiger JRP, Melcher T, Koh D-S, Sakmann B, Seeburg PH, Jonas P, Monyer H (1995) Relative abundance of subunit mRNAs determines gating and Ca^{2+} permeability of AMPA receptors in principal neurons and interneurons in rat CNS. Neuron 15:193–204

Hollmann M, Heinemann S (1994) Cloned glutamate receptors. Annu Rev Neurosci 17:31–108

Hollmann M, Maron C, Heinemann S (1994) N-glycosylation site tagging suggests a three transmembrane domain topology for the glutamate receptor GluR1. Neuron 13:1331–1343

Jonas P, Burnashev N (1995) Mechanisms controlling calcium entry through AMPA-type glutamate receptor channels. Neuron 15:987–990

Kamboj SK, Swanson GT, Cull-Candy SG (1995) Intracellular spermine confers rectification on rat calcium-permeable AMPA and kainate receptors. J Physiol 486:297–303

Koh D-S, Burnashev N, Jonas P (1995a) Block of native Ca^{2+}-permeable AMPA receptors in rat brain by intracellular polyamines generates double rectification. J Physiol 486:305–312

Koh D-S, Geiger JRP, Jonas P, Sakmann B (1995b) Ca^{2+}-permeable AMPA and NMDA receptor channels in basket cells of rat hippocampal dentate gyrus. J Physiol 485:383–402

Köhler M, Burnashev N, Sakmann B, Seeburg PH (1993) Determinants of Ca^{2+} permeability in both TM1 and TM2 of high-affinity kainate receptor channels: diversity by RNA editing. Neuron 10:491–500

Kuner T, Wollmuth LP, Karlin A, Seeburg PH, Sakmann B (1996) Structure of the NMDA receptor channel M2 segment inferred from the accessibility of substituted cysteines. Neuron 17:343–352

Kyrozis A, Goldstein PA, Heath MJS, MacDermott AB (1995) Calcium entry through a subpopulation of AMPA receptors desensitized neighboring NMDA receptors in rat dorsal horn neurons. J Physiol 485:373–381

Legendre P, Rosenmund C, Westbrook GL (1993) Inactivation of NMDA channels in cultured hippocampal neurons by intracellular calcium. J Neurosci 13:674–684

Medina I, Filippova N, Barbin G, Ben-Ari Y, Bregestovski P (1994) Kainate-induced inactivation of NMDA currents via an elevation of intracellular Ca^{2+} in hippocampal neurons. J Neurophysiol 72:456–465

Müller T, Möller T, Berger T, Schnitzer J, Kettenmann H (1992) Calcium entry through kainate receptors and resulting potassium-channel blockade in Bergmann glial cells. Science 256:1563–1566

Rozov A, Bregestovski P, Kuner T, Seeburg PH, Burnashev N (1997a) Fast interaction between AMPA and NMDA receptor channels by intracellular calcium (abstract). Society for Neuroscience Abstracts 23:1755

Rozov A, Zilberter Y, Wollmuth LP, Burnashev N (1997b) Relief from intracellular polyamine block underlies activity-dependent facilitation of currents through Ca^{2+}-permeable AMPAR channels (abstract). J Physiol 501:18P

Sommer B, Köhler M, Sprengel R, Seeburg PH (1991) RNA editing in brain controls a determinant of ion flow in glutamate-gated channels. Cell 67:11–19

Stern P, Edwards FA, Sakmann B (1992) Fast and slow components of unitary EPSCs on stellate cells elicited by focal stimulation in slices of rat visual cortex. J Physiol 449:247–278

Viklicky L Jr (1993) Calcium-mediated modulation of N-methyl-D-aspartate (NMDA) responses in cultured rat hippocampal neurones. J Physiol 470:575–600

Wood MW, VanDongen HMA, VanDongen AMJ (1995) Structural conservation of ion conduction pathways in K channels and glutamate receptors. Proc Natl Acad Sci USA 92:4882–4886

9 Localisation and Surface Expression of the AMPA Receptor Subunit GluR1 Using Green Fluorescent Protein and Anti-Peptide Antibodies

A. Doherty, A. Irving, G.L. Collingridge and J.M. Henley

9.1 Introduction

L-Glutamate is the major excitatory neurotransmitter in the vertebrate central nervous system (CNS). Glutamate receptors have been shown to be important in processes as varied as the control of motor, cardiovascular and visual functions; synapse formation, stabilisation and elimination; learning and memory; and epilepsy, neuronal degenerative disorders and the mediation neuronal cell death (for recent reviews see Bliss and Collingridge 1993; Meldrum 1994; Fletcher and Lodge 1996). In general, glutamate receptors can be characterised into two broad groups: (1) The ionotropic glutamate receptors (GluRs) which contain an integral ion channel, and (2) the metabotropic glutamate receptors (mGluRs) which are G-protein coupled.

9.1.1 Ionotropic Glutamate Receptors

Three main classes of ionotropic glutamate receptors are present in the mammalian CNS, namely the α-amino-3-hydroxyl-5-methyl-4-isoxazole propionate (AMPA), kainate and *N*-methyl-D-aspartate (NMDA) types (for extensive reviews see Gasic and Hollmann 1992; Nakanishi 1992; Seeburg 1993; Wisden and Seeburg 1993; Hollmann and Heinemann 1994).

9.1.1.1 AMPA Receptors

AMPA receptors are composed of assemblies of subunits GluR1–4 (Bettler and Mulle 1995), of which the GluR2 subunit is of particular interest because its presence, in its edited form, in the heteromeric receptor complex confers a low Ca^{2+} permeability of the ion channel (Burnashev et al. 1992). GluR1–4 each comprise approximately 900 amino acids and have calculated molecular weights (Mr) of 95–105 kDa. At the amino acid level they show 70% identity and are 20%–40% homologous to other glutamate receptor subunits (Fletcher and Lodge 1996).

9.1.1.2 Kainate Receptors

Kainate receptors are made up of combinations of GluR5–7 and KA1-KA2 subunits (Bettler and Mulle 1995). These are also approximately 900 residues long and have similar Mr values to the AMPA receptor subunits (Henley 1994a). The kainate receptor subunits share 70%–80% amino acid identity and show 35%–45% homology with the other GluR subunits.

9.1.1.3 NMDA Receptors

NMDA receptors (NMDARs) comprise NMDAR1 subunits complexed with various combinations of NMDAR2 subunits (Kutsuwada et al. 1992). Eight splice variants of the approximately 900-amino acid residue NMDAR1 subunit have been identified (Sugihara et al. 1992). Four NMDAR2 subunit clones (NMDAR2A–D) have been characterised, each of which is approximately 1500 amino acids long and are thus significantly bigger than any of the GluR subunits. NMDAR2 subunits show less than 20% sequence homology with NMDAR1 and have 55%–75% identity with each other.

9.1.2 Subunit Topology

The determination of the topology is important with respect to the design of antibodies which recognise extracellular epitopes and may, therefore, prove useful for mapping receptor distribution in living cells (Fig. 1).

Several approaches have been utilised to study the transmembrane topology of native or recombinant glutamate receptors. These include: immunocytochemical detection of regions accessible to antibodies (Petralia and Wenthold 1992; Molnar et al. 1993, 1994); protection of segments of the peptide from degradation or modification by added reagents (Wo and Oswald 1994; Bennett and Dingledine 1995); the analysis of the post-translational modification of specific sites on the protein (Raymond et al. 1993; Tingley et al. 1993; Hollmann et al. 1994; Roche et al. 1994; Taverna et al. 1994); functional analysis of mutagenised receptor (Sommer et al. 1990; Stern-Bach et al. 1994), and cell-free translation in rabbit reticulocyte lysate supplemented with canine pancreatic microsomal membranes (Seal et al. 1995).

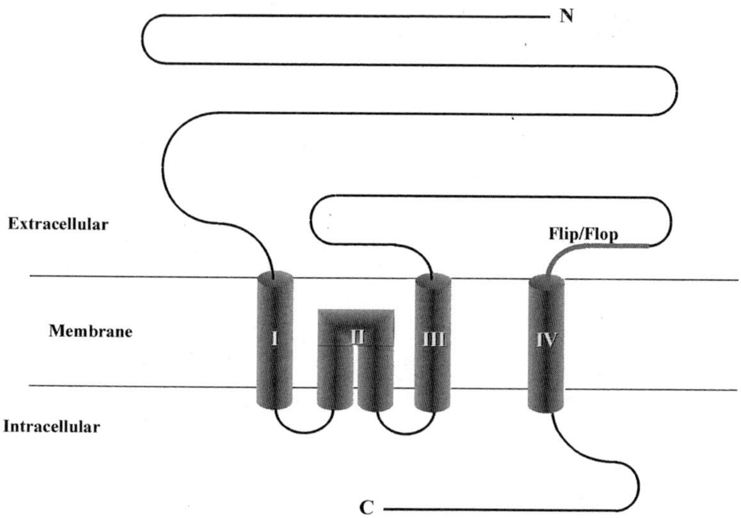

Fig. 1. Topology of an α-amino-3-hydroxy-5-methyl-4-isoxazolepropionate (AMPA) receptor subunit. The consensus model for the GluR1–4 subunits. The predicted hydrophobic regions are labelled I–IV, with region II forming an intramembrane loop. N- and C-termini are located in the extra- and intracellular domains respectively and the alternatively spliced sequence flip/flop is indicated

Initially, ionotropic glutamate receptors had been thought to belong to the ligand-gated ion channel receptor superfamily which includes nicotinic acetylcholine, γ-aminobutyric acid $(GABA)_A$, glycine and certain serotonin receptors. Based on this assumption, the two original models of topology for AMPA receptors both included four transmembrane domains (TMs) with both N and C-termini exposed on the extracellular surface (Hollmann et al. 1989; Keinanen et al. 1990). However, subsequent experimental evidence has now led to a consensus topological model which incorporates only three truly membrane spanning domains resulting in the N-terminus placed extracellularly and the C-terminus intracellularly (for review see Bettler and Mulle 1995). Thus, as illustrated in Fig. 1, although at least four membrane-associated hydrophobic domains are present, the second (TM2) domain does not traverse the membrane but rather enters the cytoplasmic face and then doubles back on itself to exit on the same (intracellular) side as it entered (Bennett and Dingledine 1995; Wo and Oswald 1995). Based on sequence analysis, a modular design has been proposed for ionotropic glutamate receptors in which the functional domains are related to prokaryotic precursor proteins (Ohara et al. 1993; Sutcliffe et al. 1996).

9.2 Reasons for Studying the Localisation and Surface Expression of AMPA Receptors

AMPA receptors play a central role in many forms of synaptic plasticity including long term potentiation (LTP) (Bliss and Collingridge 1993) and long term depression (LTD) (Bear and Abraham 1996). Interestingly, a number of groups have now shown that the majority of AMPA receptors exist as a cytoplasmic pool in neurones with only about 30% of the total number of receptors present in the neuronal plasma membrane (Petralia and Wenthold 1992; Eshhar et al. 1993; Martin et al. 1993; Molnar et al. 1993; Baude et al. 1994; Henley 1994). Arising from these observations, it has been proposed that the amplification in response to neurotransmitter release which occurs in LTP may be due to a redistribution of AMPA receptors from an intracellular pool to the postsynaptic plasma membrane. Such a mechanism could also underlie the recently reported rapid conversion of silent to functional synapses and the more slowly developing increase in AMPA receptor sensitivity in hippocampal slices during LTP (Davies et al. 1989; Isaac et al. 1995; Liao et al. 1995).

9.3 Experimental Approaches

A definitive demonstration of the hypothesised changes in the cell-surface expression of AMPA receptors requires reliable methods for the quantification of receptors in the various cellular compartments, especially in dendritic spines, under resting and synaptically potentiated/depressed conditions. Ideally this should be achieved in living cells where the movement of receptors could be followed in real time. As a first step towards these goals we are currently developing and adapting two non-invasive techniques: (1) The use of N-terminal, subunit specific antibody probes, and (2) green fluorescent protein (GFP)-tagging receptor subunits.

9.4 Antibodies Which Recognise Extracellular Epitopes on GluR1

We have obtained via collaboration and developed in-house a range of antibodies which interact with extracellular epitopes on GluRs and mGluRs. To study the cell surface and intracellular distribution of GluR1 we have primarily used an antibody which recognises an N-terminal extracellular epitope (amino acid residues 253–267) of the subunit (Baude et al. 1994; Molnar et al. 1994). Overall our results (Richmond et al. 1996) have been consistent with, and built upon, earlier subcellular fraction (Henley 1994) and microscopy studies (Petralia and Wenthold 1992; Wenthold et al. 1992; Craig et al. 1993; Huntley et al. 1994). As detailed below, these observations have implications for receptor localisation studies where it is the cell surface distribution rather than the total distribution which is usually of interest but the latter which has, prior to now, almost invariably been measured.

9.4.1 Cell Culture

Cultures of rat hippocampal neurones were prepared from two-day-old rat pups. The animals were sacrificed, the hippocampus removed and 500-μm slices cut. The slices were washed in Hepes buffered saline (HBS) and incubated with 0.5 mg/ml^{-1} pronase E and protease type X for 30 min. Following further washing the cells were dissociated by trituration and harvested by centrifugation at 100 *g* for 6 min. The pelleted cells were washed in HBS once more by centrifugation and then plated on poly-L-lysine-coated cover slips. Cultures were grown at 37°C in 90% MEM, 10% dialysed foetal bovine serum and 2 m*M* L-glutamine. Cytosine arabinofuranoside (5 μ*M*) was added at culture day 3 to inhibit glial cell proliferation. The cultures were maintained for up to 5 weeks.

9.4.2 Antibodies and Imaging

The generation of the GluR1 antibody has been described previously (Baude et al. 1994; Molnar et al. 1994; Richmond et al. 1996). Other antibodies were obtained from Chemicon (Harrow, UK). In all experiments each of the antibodies used was at a final concentration of 1–5 $\mu g/ml^{-1}$.

For confocal experiments on living cells the cultures were exposed to the GluR1 antibody for 40 min at room temperature, washed twice and then incubated with tetramethylrhodamine isothiocyanate (TRITC)-conjugated goat anti-rabbit secondary antibody for a further 30 min. In colocalisation experiments using antibodies requiring access to the cell interior the same protocol was used and then, after incubation and washing to remove non-bound TRITC antibody, the cells were fixed with 3% paraformaldehyde for 20 min, washed and then permeabilised with 0.1% Triton X-100 for 10 min. The cells were then reprobed as above but using 7-amino-4-methylcoumarin-3-acetic acid (AMCA)-conjugated goat anti-rabbit secondary antibody.

Imaging was done with a MRC-BioRad 600/UV laser scanning confocal microscope. TRITC was excited at 514 nm, AMCA and indo-1 were excited at 351 nm.

9.4.3 Surface Distribution of GluR1 Subunits in Living Hippocampal Neurones

On living neurones pronounced labelling with the N-terminal-directed GluR1 antibody was present on the cell body and discrete patches of labelling were observed in the dendrites (Fig. 2). To determine whether the GluR1 subunit was expressed on all cells or just on neurones, the fluorescent dye indo-1 was added to the living cultures after antibody imaging. Glial cells were immunonegative. Preincubation of the antibody with the peptide against which it was raised blocked specific staining of living neurones. No labelling was obtained with the C-terminal antibody in living cells but, in agreement with other studies (Craig et al. 1993, 1994; Eshhar et al. 1993), high levels of immunoreactivity were obtained in neurones permeabilised with 0.1% Triton X-100.

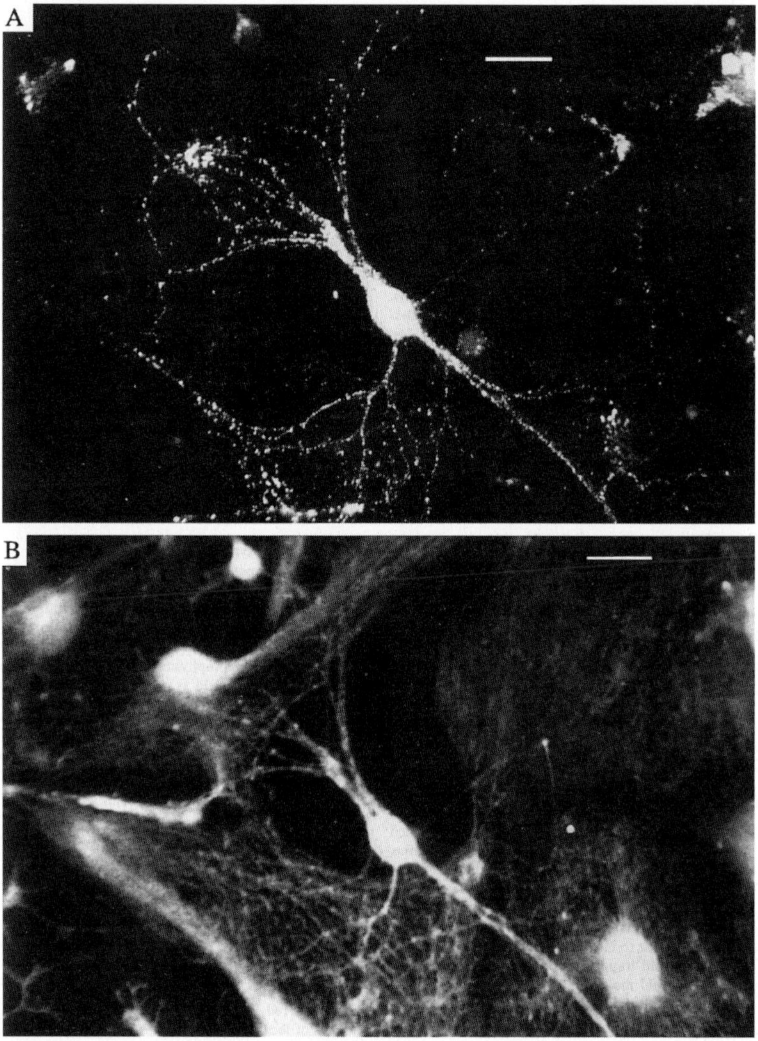

Fig. 2A,B. Surface distribution of glutamate receptor (GluR)1 subunits on living hippocampal neurones. **A** Punctate staining of 12-day-old cells labelled with the N-terminal GluR1 antibody and visualised with a tetramethylrhodamine isothiocyanate (TRITC)-conjugated goat anti-rabbit secondary antibody. **B** The same field of view following loading with indo-1-acetoxymethyl (AM) to visualise all cells. Scale bars=25 μm

9.4.4 Clusters of Surface-Expressed GluR1

Dendritic puncta of GluR1 immunoreactivity were 0.5–2 μm in diameter and on fine distal dendrites these were spaced 1–2 μm apart. Estimates for the spacing of puncta on proximal dendrites could not be made due to the high intensity of the signal. Significantly, in mature neuronal cultures the hot-spots of GluR immunoreactivity were associated with dendritic spine-like processes (spines). No significant movement of the puncta was observed over a 1-h interval.

9.4.5 Developmental Time-Course of GluR1 Immunoreactivity

The surface-expressed GluR1 immunoreactivity was not present on neurones until 3–5 days in culture. At early stages after day 5 considerable variability in the degree of staining was observed. Maximal staining was seen after 3–4 weeks in culture and maintained up to 5 weeks (the longest period tested) when all neurones were labelled. In mature cultures optical Z-sections through the neuronal cell body revealed distinct plasma membrane-associated puncta with essentially no immunostaining in the cytoplasm (Fig. 3).

9.4.6 Colocalisation of GluR1 and Synaptophysin

The proportions of surface-expressed GluR1 subunits which are synaptically or extrasynaptically localised is a fundamental issue. Therefore, we determined the extent of colocalisation of GluR1 in living cells with synaptophysin in the same cells following permeabilisation. Since pre- and postsynaptic sites were not distinguishable at the level of resolution used, synaptophysin immunoreactivity served as a marker for synapses in general (Fletcher et al. 1991; Craig et al. 1994). Synaptophysin labelling was observed in discrete puncta on the soma and dendritic processes. In young cultures (5–6 days) all GluR1 immunoreactivity colocalised with the synaptophysin label. However, many puncta of synaptophysin labelling were not labelled by the GluR1 antibody. In addition, in older cultures GluR1-positive puncta appeared that were immunonegative for synaptophysin.

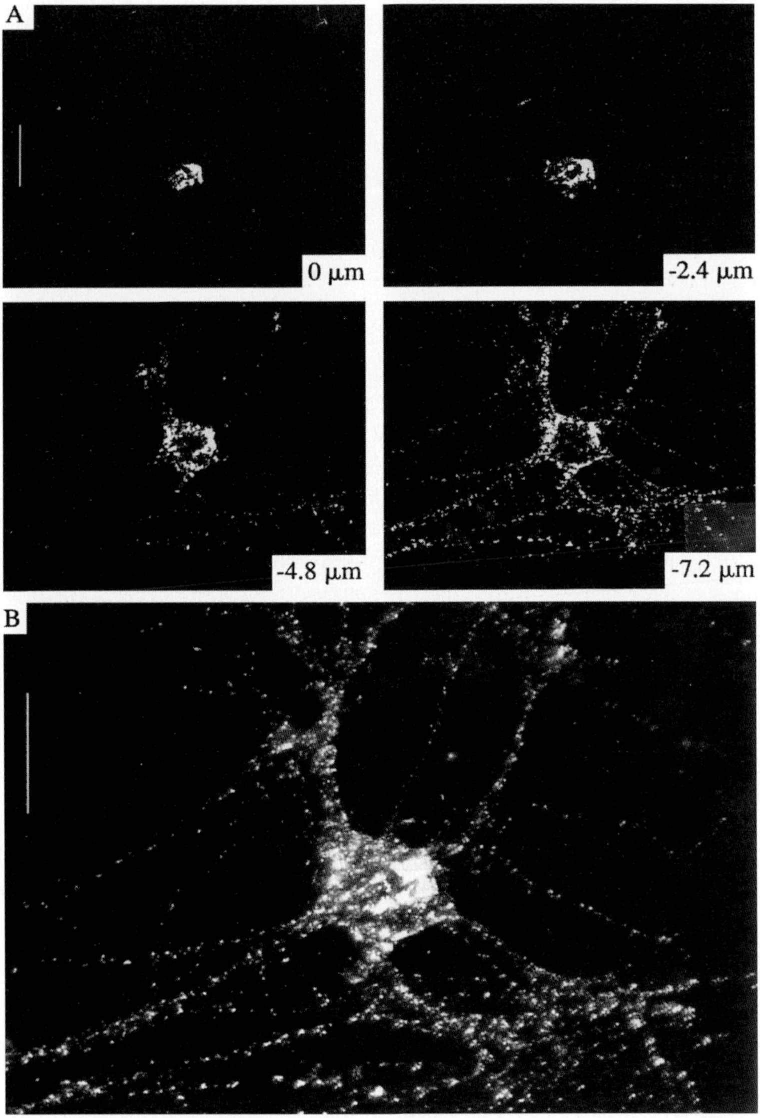

Fig. 3A,B. Legend see p. 171

9.4.7 Differences in the Distribution of GluR1 Immunoreactivity in Living and Permeabilised Neurones

Only the cell surface-expressed GluR1 subunits should be detected in living cells under the conditions used. In fixed, permeabilised cells, however, the total population of intracellular and plasma membrane-inserted receptors can be determined. To estimate the fraction of receptor subunits which are surface-expressed, living cells were exposed to the N-terminal GluR1 antibody and then, after washing, labelled with the secondary antibody. After further washing the cells were subsequently fixed and permeabilised and were exposed to either a commercially available C-terminal antibody specific for GluR1 or rechallenged with our N-terminal antibody (Fig. 4). A different secondary antibody was used to visualise the GluR1 subunit accessed following permeabilisation. Marked differences in the labelling were detected between the two conditions. While in both conditions distinct puncta were observed in the soma and dendrites, after permeabilisation diffuse intracellular immunoreactivity was evident in the cell body and in the proximal dendrites. Interestingly, some dendritic spines that were not immunopositive in living cells were labelled by the anti-GluR1 antibodies after permeabilisation.

Our colocalisation studies of cell-surface GluR1 subunits with the N-terminal GluR1 antibody in live cells and synaptophysin in fixed cells demonstrated the clusters of both synaptic and extrasynaptic GluR1 immunoreactivity. The occurrence of non-synaptic puncta of GluR1 was dependent on the time in culture. At short culture times (3–5 days) the GluR1 immunoreactivity highly correlated to synaptophysin immunoreactivity, suggesting that the subunit (and, by implication, AMPA receptors in general) occur only at points of synaptic contact. In more mature cultures extrasynaptic clusters of GluR1 immunoreactivity may reflect later developmental processes.

◀ **Fig. 3A,B.** Z-sections showing the cellular distribution of glutamate receptor (GluR)-1 in a single neurone. **A** Optical sections were taken at 2.4-µm increments through a neurone after 15 days in culture. Note the lack of intrasomal staining. **B** Montage of the four sections at higher magnification showing clustering of GluR1 on the soma and dendrites

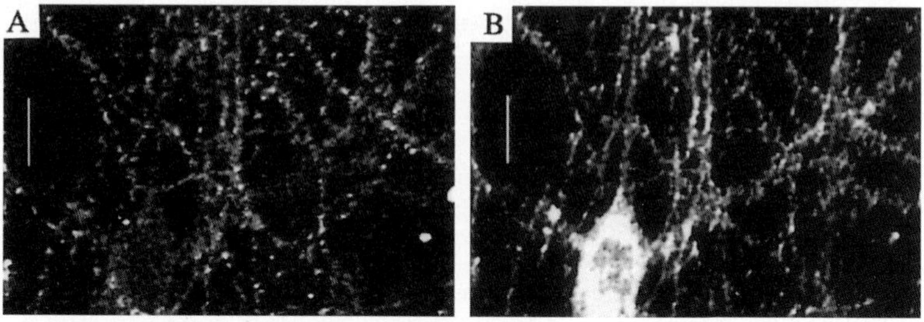

Fig. 4A,B. Cell surface and total distribution of glutamate receptor (GluR)-1. **A** N-terminal antibody immunostaining of GluR1 on an intact living hippocampal neurone after 36 days in culture. **B** Total distribution of GluR1 in the same neurone following fixation, permeabilisation and reprobing with additional N-terminal antibody

9.4.8 Intracellular GluR1

The diffuse GluR1 immunoreactivity detected inside the soma and proximal dendrites is likely to reflect trafficking and intracellular transport of the subunit/AMPA receptor complex from its site of synthesis in the cell body to its site of action at synapses located in dendritic spines. Perhaps a more interesting observation is that some synapses which were not immunoreactive in living neurones are GluR1-immunopositive on permeabilisation. As discussed above, an attractive explanation is that these synapses represent the recently reported silent synapses. Thus the presence of receptors within the dendritic spine would allow for the very rapid insertion of these receptors into the post-synaptic membrane, thereby converting an AMPA-unresponsive synapse into an AMPA-responsive synapse within seconds.

9.5 Green Fluorescent Protein-Tagged Glutamate Receptors

9.5.1 Green Fluorescent Protein

GFP is a 238-amino acid 27-kDa protein originally isolated form the jellyfish *Aequorea victoria* which provides a novel tool for monitoring gene expression and protein localisation in vivo. GFP fluorescence properties (excitation γ_{max}=395 nm with a minor peak at 470 nm and emission γ_{max}=508 nm with a shoulder at 540 nm) are intrinsic to the protein and do not require additional co-factors. Cotransfection of wild-type GFP with cloned receptors has been used as a tool to successfully identify transfected human embryonic kidney (HEK) 293 cells for subsequent electrophysiological analysis (Marshall et al. 1995). Recently, mutant forms of GFP have been engineered which exhibit much greater fluorescence than native GFP and in our labs we have used a triple-mutant GFP (GFPm; which includes a S65T mutation) which gives stronger fluorescence and is resistant to photobleaching to visualise HEK 293 cells cotransfected with NMDA receptor subunits (Bresink et al. 1996). Furthermore, GFP mutants are now being developed which display different colour emissions, for example blue fluorescent protein (BFP) (Heim et al. 1994; Heim and Tsien 1996). BFP has a single amino acid substitution of histidine for tyrosine at position 66 (Y66H-GFP) which results in a spectral shift with an excitation γ_{max}=382 nm and an emission γ_{max}=448 nm giving blue fluorescence). These brighter mutant GFPs with better excitation/emission profiles provide considerable advantages and the possibility of multicolour applications in which differential gene expression and intracellular trafficking can be examined (Rizzuto et al. 1996). For example, a recent report has demonstrated simultaneous analysis and sorting of wild-type GFP- and BFP-expressing populations of HEK cells by flow cytometry (Ropp et al. 1996).

9.5.2 Receptor-GFP Fusion Proteins

9.5.2.1 NMDAR1-GFP Chimera

Initial studies from other research groups have demonstrated the feasibility of using GFP-fusion constructs to monitor protein trafficking. For example, wild-type GFP has been fused to the 3' end of the NMDAR1 subunit and the resulting construct expressed in HEK cells (Marshall et al. 1995). In contrast to the signal observed with GFP alone, the NMDAR1-GFP chimera had a more restricted distribution within the cell with evidence of compartmentalisation at the plasma membrane and in perinuclear organelles. Cotransfection of the GFP-NMDAR1 chimera with cDNAs encoding NMDAR2 subunits resulted in the expression of functional NMDA receptors. The pharmacological and biophysical properties if the receptors containing NMDAR1-GFP were similar to recombinant receptors containing NMDAR1 subunits not fused to GFP (Marshall et al. 1995).

9.5.2.2 Transporter Protein-GFP Chimera

Similarly, GFP has been used to tag an insulin-regulated glucose transporter GLUT4. Exposure to insulin induces the translocation of GLUT4 from an intracellular location to the cell membrane. Chimeric constructs have been engineered with GFP at either the N-terminal or C-terminal end of GLUT4 (GFP-GLUT4 and GLUT4-GFP, respectively). Expression in CHO cells resulted in a localised cytoplasmic distribution of GFP fluorescence. Insulin stimulation resulted in a net increase in plasma membrane fluorescence. GLUT4-GFP but not GFP-GLUT4 was re-internalised following insulin removal suggesting that specific internalisation signals at the N-terminal of GLUT4 are disrupted by fusion with GFP (Dobson et al. 1996).

9.5.2.3 G-Protein-Coupled Receptor-GFP Chimera

Recently, the pharmacological properties, physiological responses, internal trafficking and surface mobility of a β_2-adrenergic receptor (β_2AR)–mutant GFP (S65T/GFP) conjugate expressed in HEK 293 cells has been reported (Barak et al. 1997). The β_2AR–S65T/GFP chimera shows strong, diffuse plasma membrane fluorescence when expressed in HEK cells. It binds agonist and antagonist, stimulates adenylyl cyclase, undergoes phosphorylation and is internalised in a manner indistinguishable from wild-type receptors. The intracellular trafficking and surface mobility were readily observable using the GFP fluorescence (Barak et al. 1997).

9.5.3 Construction of GluR1-GFP Fusion Proteins

We have constructed a series of glutamate receptor subunit-GFP and subunit-GFPm fusion proteins, including both a truncated and, more recently, a full-length GluR1 tagged at the C-terminal end with GFPm. The behaviour and characteristics of these fusions in HEK 293 cells are currently under investigation.

9.5.3.1 Truncated GluR1(GluR1/t)-GFP Fusion Protein

GluR1(flop) was cloned into the cytomegalovirus (CMV)-driven expression vector pcDNA1/amp (Invitrogen) from pBlueScript KS+ and digested using AccIII and XhoI, removing all the GluR1 coding sequence from just upstream of the third hydrophobic region. GFPm, cloned into another CMV driven expression vector, pCMX, was amplified using the primers 5'-CGG ACT CCG GAT CCA AGC TTC TAG-3' (forward) and 5'-TGG CCT CGA GCC TCT AGA TAC CTA-3'(reverse), incorporating an AccIII site on the 5' end and a STOP codon and XhoI site on the 3' end of the polymerase chain reaction (PCR) fragment. Following digestion, the GFPm fragment was used to replace the deleted GluR1 sequence, giving a fusion protein consisting of the N-terminus and the first two hydrophobic domains of GluR1 and GFPm.

9.5.3.2 Full-Length GluR1-GFP Fusion Protein

The full-length GluR1-GFP fusion construction was carried out in two parts. The C-terminus of the GluR1 coding sequence cloned into pcDNA1/amp was first amplified from a BalI site within the fourth hydrophobic domain, using primers 5'- CTG GGA CTG GCC ATG CTG GTT-3' (forward) and 5'- G ACG TCT CGA GGA ACT AAG CTT CAA TCC TGT GGC TCC-3' (reverse). This removed the STOP codon and 3' untranslated region and inserted a HindIII site at the 3' end of the GluR1 sequence. The mutagenesis and fidelity of the amplification was confirmed by direct sequencing. The resulting sequence (GluR1-STOP) was then subcloned into pCMX containing GFPm using HindIII, which gave a full-length fusion with a five-amino-acid linker between the two proteins.

9.5.3.3 Expression of Fusion Proteins

Native GluR1, GFPm, GluR1(t)-GFPm and GluR1-GFPm were expressed in HEK 293 cells. Cells were grown on glass coverslips (22-mm diameter, 0 thickness) and transfected with 10 μg of each DNA per coverslip by calcium phosphate precipitation as previously described (Bresink et al. 1996). Transfection efficiency was 30%–40% and cells were used 48–72 h after transfection.

9.5.3.4 Immunocytochemistry and Western Blotting

For immunoblots, cells expressing the appropriate cDNA were solubilised in sodium dodecyl sulfate (SDS) buffer and the protein precipitated using chloroform and methanol (Wessel and Flügge 1984). A total of 20 μg of protein was used per lane. For immunostaining, cells were washed twice in phosphate-buffered solution (PBS) and used either live or after fixation with methanol at –20°C for 10 min. Primary antibody incubations (90 min at 22°C) were carried out in PBS containing 10 mg/ml^{-1} bovine serum albumin (BSA) and 1 g/l^{-1} D-glucose to reduce non-specific staining and to help maintain live cells. A rhodamine-conjugated secondary antibody was used to allow immunostaining to be distinguished from the GFP fluorescence.

9.5.4 Imaging of GFP and GluR1-GFP in Cells

Imaging was carried out on a Bio-Rad MRC1024 laser scanning confocal microscope equipped with an argon ion laser. GFPm fluorescence was detected using 488 nm and 515 nm excitation and emission filters, while rhodamine fluorescence was detected using 514 nm and 585 nm long pass filters. Under these conditions, no GFPm fluorescence was detected in the rhodamine channel. In all the images shown, the field of view is filled with cells.

9.5.5 Expression of GluR1(t)-GFP in HEK 293 Cells

Following construction of the fusion proteins, GluR1$_{flop}$, GFPm and GluR1(t)-GFPm were expressed in HEK 293 cells. Western blots performed using polyclonal antibodies directed against the C-terminus of GluR1 (1 μg/ml^{-1}; Chemicon) and GFP (1:1000; Clontech). The anti-GluR1 antibody recognised a species of 100 kDa in cells expressing native GluR1(flop), but not in cells expressing either GFPm or GluR1(t)-GFPm (see Fig. 7). Conversely, the anti-GFP antibody recognised species of about 100 kDa and 30 kDa in cells expressing GluR1(t)-GFPm and GFPm, respectively, but nothing in cells expressing GluR1$_{flop}$. No immunoreactive species were detected with either antibody in untransfected HEK 293 cells. The sizes of the respective

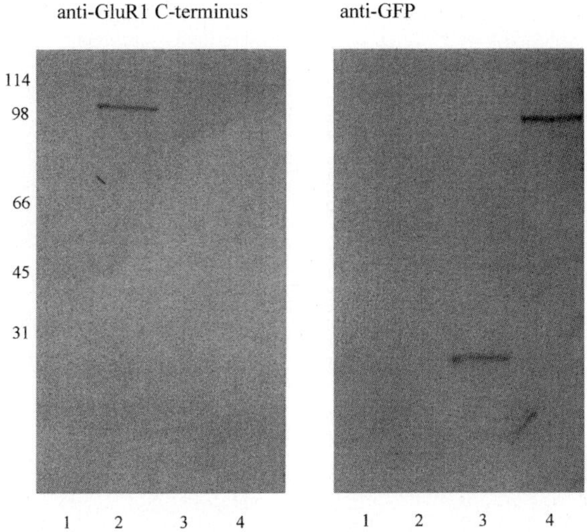

Fig. 5. Western blots demonstrating expression of a partial length glutamate receptor (GluR)-1(t)-GFPm fusion protein in human embryonic kidney (HEK) 293 cells. Lanes on each blot represent 20 μg protein from cells transfected with: (*1*) no DNA, (*2*) GluR1$_{\text{flop}}$, (*3*) GFPm and (*4*) GluR1(t)-GFPm. The blots were probed with either anti-GluR1 C-terminal antibody or anti-GFP antibody as indicated

immunoreactive species detected are consistent with the expected sizes of GluR1, GFPm and GluR1(t)-GFPm (Fig. 5). These results demonstrate the replacement of the C-terminal portion of GluR1(flop) with GFPm.

9.5.5.1 Different Distributions of GFPm and Subunit-GFPm Fusion Proteins

The distribution of GFP fluorescence was investigated in both live and fixed HEK 293 cells expressing either GFPm of GluR1(t)-GFPm. Following transfection, cells were viewed by confocal microscopy under 488-nm excitation. In live cells, GFPm showed a uniform fluorescence throughout the cell including the nucleus. GluR1(t)-GFPm, however, showed a much more localised distribution. In many cells intense staining was seen adjacent to the nucleus and in halo around the edge of the

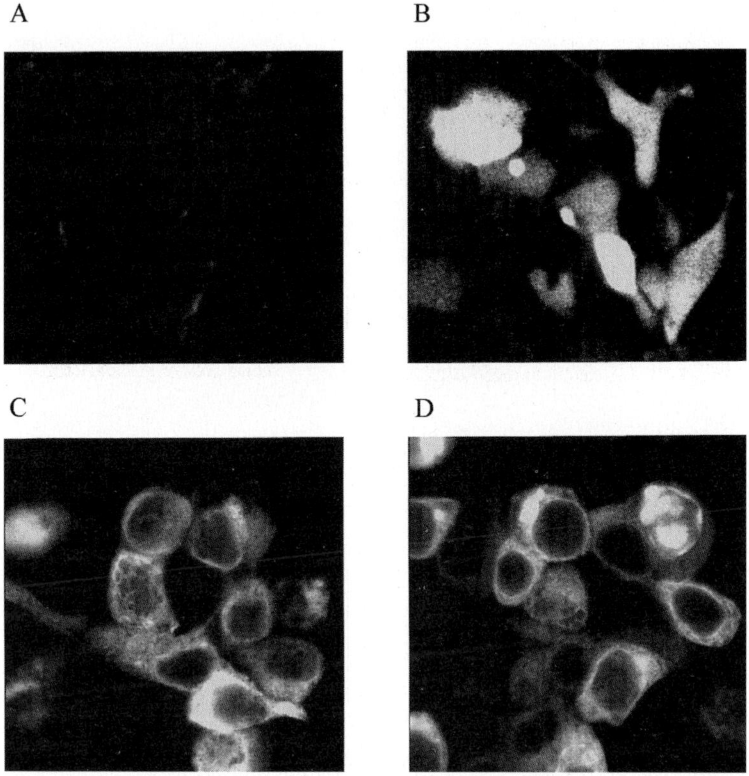

Fig. 6A–D. Visualisation of human embryonic kidney (HEK) 293 cells expressing glutamate receptor (GluR)-1–triple-mutant green fluorescent protein (GFPm) fusion proteins. Cells were transfected and viewed 48 h later under 488/515 nm excitation/emission, respectively. Fluorescence from cells transfected with: **A** untransfected, **B** GFPm alone, **C** truncated GluR1(t)-GFPm, and **D** full-length GluR1-GFPm

cell, suggesting an association of GluR1(t)-GFPm with the plasma membrane. No intranuclear staining was seen (Fig. 6). This fluorescence pattern was maintained in cells fixed with methanol. In contrast, GFPm fluorescence was lost following fixation (data not shown), suggesting that GluR1(t)-GFPm is associated with membranes while GFPm itself remains in a soluble compartment. We have recently obtained

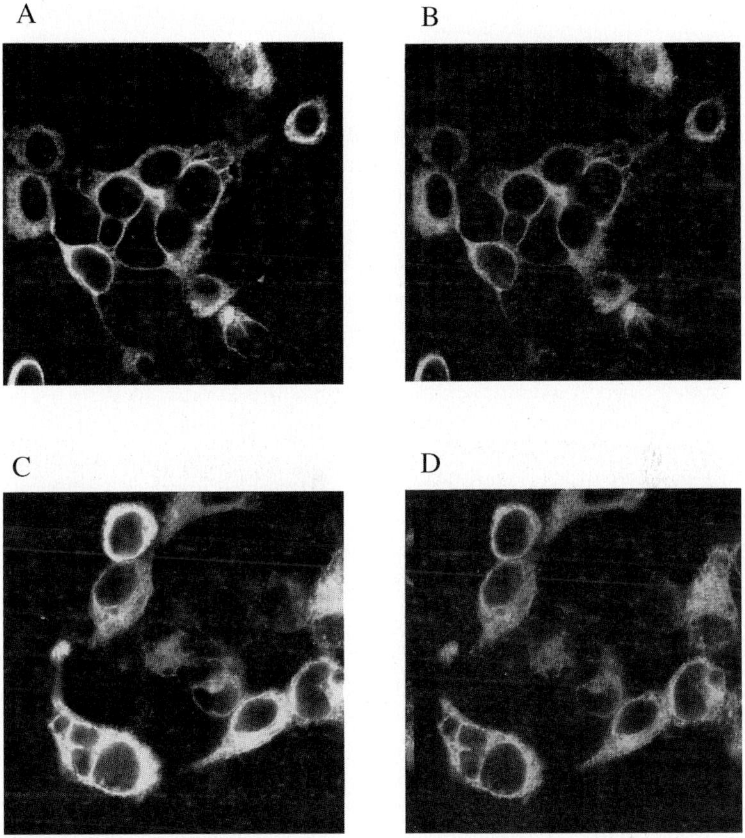

Fig. 7A–D. Colocalisation of mutant green fluorescent protein (GFPm) fluorescence and GFPm and glutamate receptor (GluR)-1 immunoreactivity. Human embryonic kidney (HEK) 293 cells expressing GluR1(t)-GFPm were fixed with methanol at –20°C and probed with antibodies directed against the GluR1 N-terminus or GFPm as indicated. Immunostaining was visualised with rhodamine-conjugated goat anti-rabbit secondary antibody at 514/585 nm excitation/emission wavelengths, respectively. GFPm was visualised at 488/515 nm excitation/emission wavelengths, respectively. No GFPm fluorescence was detected at the rhodamine wavelengths. **A** GFPm fluorescence, **B** GFPm immunostaining in the same field of view as (**A**), **C** GFPm fluorescence, **D** GluR1 immunostaining in the same field of view as (**C**)

essentially identical distributions with the full-length GluR1-GFPm fusion.

In order to demonstrate that the fluorescence seen under 488-nm illumination was due to the GluR1(t)-GFPm fusion, HEK 293 cells transfected with this construct were fixed with methanol, probed with the anti-GFP antibody (1:200) and immunoreactivity detected using 514-nm excitation . There is an absolute colocalisation of the GFP fluorescence and anti-GFP immunoreactivity (Fig. 7). This is further supported by similar experiments using our polyclonal antibody directed against the N-terminus of GluR1. HEK 293 cells were transfected with GluR1(t)-GFPm, fixed in methanol and probed with the anti-N-terminal GluR1 antibody (1 $\mu g/ml^{-1}$). As with the anti-GFP antibody, immunoreactivity was colocalised with the GFP fluorescence.

These results demonstrate the construction and distribution of a truncated GluR1(t)-GFPm and full-length GluR1-GFPm fusion proteins in a mammalian cell line. The distribution pattern is consistent with the targeting of subunit-GFPm fusions to perinuclear and plasma membrane domains. A similarly truncated, but not GFP-tagged, GluR1 subunit has recently been shown to be incorporated into the plasma membrane of Cos-7 cells (McIlhinney and Molnar 1996). However, we have thus far been unable to demonstrate consistently cell surface immunostaining of live HEK 293 cells transfected with our wild-type GluR1 (kindly supplied by S. Heinemann and J. Boulter). At present we suspect that there is a problem with the 5' untranslated region of the cDNA which results in the incorrect or impaired surface expression of the receptor subunit. We are currently investigating this possibility and also testing the equivalent GluRA clone (kindly provided by P. Seeburg).

9.5.6 Expression of Receptor-GFPm Fusion Proteins in Neurones

We are investigating adenovirus transfection methods as a means of expressing the GFP-tagged receptors in neuronal cell cultures. We have shown that the adenovirus successfully transfects our hippocampal cultures using the *lac z* and GFP reporter genes. Present work is aimed towards using adenoviruses which are neuronally specific and contain a tetracycline-inducible promoter. The neuronal specificity avoids problems with high background due to glial cell expression and the incorporation of an inducible promoter will allow expression levels to be regulated, thereby minimising cell toxicity and the misprocessing and mistargeting events that can be associated with protein overexpression. Once the surface expression difficulties with the GluR1 subunit have been resolved we shall incorporate the appropriate GFPm-tagged subunit into neurones.

9.6 Concluding Remarks

Each of the approaches detailed above has inherent advantages and disadvantages. However, our eventual aim is to combine the use of antibodies recognising extracellular epitopes with the expression of GFPm-tagged receptor subunits in neuronal cells as a means of monitoring the cell biology of glutamate receptors. We believe that this strategy has the potential to provide a powerful tool for the analysis of the dynamics of GluR movements and surface expression in living cultured neurones under physiological and pathological conditions.

References

Barak LS, Ferguson SSG, Zhang J, Martinson C, Meyer, T and Caron, MG (1997) Internal trafficking and surface mobility of a functionally intact β_2-adrenergic receptor-green fluorescent protein conjugate. Mol Pharmacol 51:177–184

Baude A, Molnar E, Latawiec D, Mcilhinney RAJ, Somogyi P (1994) Synaptic and nonsynaptic localization of the GluR1 subunit of the AMPA-type excitatory amino acid receptor in the rat cerebellum. J Neurosci 14:2830–2843

Bear MF, Abraham WC (1996) Long-term depression in the hippocampus. Annu Rev Neurosci 19:437–462

Bennett JA, Dingledine R (1995) Topology profile for a glutamate receptor: three transmembrane domains and a channel-lining reentrant membrane loop. Neuron 14:373–384

Bettler B, Mulle C (1995) AMPA and kainate receptors. Neuropharmacology 34:123–140

Bliss T, Collingridge G (1993) A synaptic model of memory: long-term potentiation in the hippocampus. Nature 361:31–39

Bresink I, Benke TA, Collett VJ, Seal AJ, Parsons CG, Henley JM, Collingridge GL (1996) Effects of memantine and (+)-MK-801 on recombinant rat NMDA receptors expressed in HEK 293 cells. Br J Pharmacol 119:195–204

Burnashev N, Monyer H, Seeburg PH, Sakmann B (1992) Divalent ion permeability of AMPA receptor channels is dominated by the edited form of a single subunit. Neuron 8:189–198

Craig AM, Blackstone CD, Huganir RL, Banker G (1993) The distribution of glutamate receptors in cultured rat hippocampal neurons – postsynaptic clustering of AMPA-selective subunits. Neuron 10:1055–1068

Craig AM, Blackstone CD, Huganir RL, Banker G (1994) Selective clustering of glutamate and gamma-aminobutyric acid receptors opposite terminals releasing the corresponding neurotransmitters. Proc Natl Acad Sci USA 91:12373–12377

Davies SN, Lester RAJ, Reymann KG, Collingridge GL (1989) Temporally distinct pre- and post-synaptic mechanisms maintain long-term potentiation. Nature 338:500–503

Dobson SP, Livingstone C, Gould GW, Tavaré JM (1996) Dynamics of insulin-stimulated translocation of GLUT4 in single living cells visualised using green fluorescent protein. FEBS Lett 393:179–184

Eshhar N, Petralia RS, Winters CA, Niedzielski AS, Wenthold RJ (1993) The segregation and expression of glutamate receptor subunits in cultured hippocampal neurons. Neuroscience 57:943–964

Fletcher EJ, Lodge D (1996) New developments in the molecular pharmacology of AMPA and kainate receptors. Pharmacol Ther 70:65–89

Fletcher TL, Cameron P, De Camilli P, Banker G (1991) The distribution of synapsin 1 and synaptophysin in hippocampal neurons developing in culture. J Neurosci 1617–1626

Gasic GP, Hollmann M (1992) Molecular neurobiology of glutamate receptors. Annu Rev Physiol 54:507–36

Heim R, Tsien RY (1996) Engineering green fluorescent protein for improved brightness, longer wavelengths and fluorescence resonance energy transfer. Curr Biol 6:178–182

Heim R, Prasher DC, Tsien RY (1994) Wavelength mutations and posttranslational autoxidation of green fluorescent protein. Proc Natl Acad Sci USA 91:12501–12504

Henley JM (1994a) Kainate-binding proteins: phylogeny, structures and possible functions. Trends Pharmacol Sci 15:182–190

Henley JM (1994b) Subcellular localisation of AMPA receptors in rat hippocampus. Biochem Soc Trans 22:227S

Hollmann M, Heinemann S (1994) Cloned glutamate receptors. Annu Rev Neurosci 17:31–108

Hollmann M, O'Shea-Greenfield A, Rogers SW, Heinemann S (1989) Cloning by functional expression of a member of the glutamate receptor family. Nature 342:643–648
Hollmann M, Maron C, Heinemann S (1994) N-glycosylation site tagging suggests a three transmembrane domain topology for the glutamate receptor GluR1. Neuron 13:1331–1343
Huntley GW, Vickers JC, Janssen W, Brose N, Heinemann SF, Morrison JH (1994) Distribution and synaptic localization of immunocytochemically identified NMDA receptor subunit proteins in sensory-motor and visual cortices of monkey and human. J Neurosci 14:3603–3619
Isaac JTR, Nicoll RA, Malenka RC (1995) Evidence for silent synapses: implications for expression of LTP. Neuron 15:427–434
Keinanen K, Wisden W, Sommer B, Werner P, Herb A, Verdoorn TA, Sakmann B, Seeburg PH (1990) A family of AMPA-selective glutamate receptors. Science 249:556–560
Kutsuwada T, Kashiwabuchi N, Mori H, Sakimura K, Kushiya E, Araki K, Meguro H, Masaki H, Kumanishi T, Arakawa M, Mishina M (1992) Molecular diversity of the NMDA receptor channel. Nature 358:36–41
Liao D, Hessler NA, Malinow R (1995) Activation of postsynaptically silent synapses during pairing-induced LTP in CA1 region of hippocampal slice. Nature 375:400–404
Marshall J, Molloy R, Moss GWJ, Howe JR, Hughes TE (1995) The jellyfish green fluorescent protein: a new tool for studying ion channel expression and function. Neuron 14:211–215
Martin LJ, Blackstone CD, Levey AI, Huganir RL, Price DL (1993) AMPA glutamate receptor subunits are differentially distributed in rat brain. Neuroscience 53:327–358
McIlhinney RAJ, Molnar E (1996) Characterisation, cell-surface expression and ligand-binding properties of different truncated N-terminal extracellular domains of the ionotropic glutamate receptor subunit GluR1. Biochem J 315:217–225
Meldrum BS (1994) The role of glutamate in epilepsy and other CNS disorders. Neurology 44:14–23
Molnar E, Baude A, Richmond SA, Patel PB, Somogyi P, Mcilhinney RAJ (1993) Biochemical and immunocytochemical characterization of antipeptide antibodies to a cloned GluR1 glutamate receptor subunit – cellular and subcellular distribution in the rat forebrain. Neuroscience 53:307–326
Molnar E, Mcilhinney RAJ, Baude A, Nusser Z, Somogyi P (1994) Membrane topology of the GluR1 glutamate receptor subunit: epitope mapping by site-directed antipeptide antibodies. J Neurochem 63:683–693
Nakanishi S (1992) Molecular diversity of glutamate receptors and implications for brain function. Science 258:597–603

Ohara PJ, Sheppard PO, Thogersen H, Venezia D, Haldeman BA, Mcgrane V, Houamed KM, Thomsen C, Gilbert TL, Mulvihill ER (1993) The ligand-binding domain in metabotropic glutamate receptors is related to bacterial periplasmic binding proteins. Neuron 11:41–52

Petralia RS, Wenthold RJ (1992) Light and electron immunocytochemical localization of AMPA-selective glutamate receptors in the rat brain. J Comp Neurol 318:329–354

Raymond LA, Blackstone CD, Huganir RL (1993) Phosphorylation and modulation of recombinant GluR6 glutamate receptors by cAMP-dependent protein kinase. Nature 361:637–641

Richmond SA, Irving AJ, Molnar E, McIlhinney RAJ, Michelangelli F, Henley JM, Collingridge GL (1996) Localisation of the glutamate receptor subunit GluR1 on the surface of living and within cultured hippocampal neurones. Neuroscience 75:69–82

Rizzuto R, Brini M, De Giorgi F, Rossi R, Heim R, Tsien RY, Pozzan T (1996) Double labelling of subcellular structures with organelle-targeted GFP mutants in vivo. Curr Biol 6:183–186

Roche KW, Raymond LA, Blackstone C, Huganir RL (1994) Transmembrane topology of the glutamate receptor subunit glur6. J Biol Chem 269:11679–11682

Ropp JD, Donahue CJ, Wolfgang-Kimball D, Hooley JJ, Chin JYW, Cuthbertson RA, Bauer KD (1996) Aequorea green flourescent protein: simultaneous analysis of wild-type and blue-fluorescing mutanat by flow cytometry. Cytometry 24:284–288

Seal AJ, Collingridge GL, Henley JM (1995) An investigation of the membrane topology of the ionotropic glutamate receptor subunit GluR1 in a cell-free system. Biochem J 312:451–456

Seeburg PH (1993) The TiPS/TINS lecture: the molecular biology of mammalian glutamate receptor channels. Trends Pharmacol Sci 14:297–303

Sommer B, Keinanen K, Verdoorn TA, Wisden W, Burnashev N, Herb A, Kohler M, Takagi T, Sakmann B, Seeburg PH (1990) Flip and flop – a cell-specific functional switch in glutamate-operated channels of the CNS. Science 249:1580–1585

Stern-Bach Y, Bettler B, Hartley M, Sheppard PO, O'Hara PJ, Heinemann SF (1994) Agonist selectivity of glutamate receptors is specified by two domains structurally related to bacterial amino acid-binding proteins. Neuron 13:1345–1357

Sugihara H, Moriyoshi K, Ishii T, Masu M, Nakanishi S (1992) Structures and properties of 7 isoforms of the NMDA receptor generated by alternative splicing. Biochem Biophys Res Commun 185:826–832

Sutcliffe MJ, Wo ZG, Oswald Re (1996) Three-dimensional models of non-NMDA glutamate receptors. Biophys J 70:1575–1589

Taverna FA, Wang LY, Macdonald JF, Hampson DR (1994) A transmembrane model for an ionotropic glutamate receptor predicted on the basis of the location of asparagine-linked oligosaccharides. J Biol Chem 269:14159–14164

Tingley WG, Roche KW, Thompson AK, Huganir RL (1993) Regulation of NMDA receptor phosphorylation by alternative splicing of the C-terminal domain. Nature 364:70–73

Wenthold RJ, Yokotani N, Doi K, Wada K (1992) Immunochemical characterisation of the non-NMDA glutamate receptor using subunit specific antibodies. J Biol Chem 267:501–507

Wessel D, Flügge UI (1984) A method for the quantitative recovery of protein in dilute solution in the presence of detergents and lipids. Anal Biochem 138:141–143

Wisden W, Seeburg PH (1993) Mammalian ionotropic glutamate receptors. Curr Opin Neurobiol 3:291–298

Wo ZG, Oswald RE (1994) Transmembrane topology of two kainate receptor subunits revealed by N-glycosylation. Proc Natl Acad Sci USA 91:7154–7158

Wo Z, Oswald RE (1995) Unravelling the modular design of glutamate-gated ion channels. Trends Neurosci 18:161–168

10 Input- and Output-Specific Segregation of Amino Acid Neurotransmitter Receptors on the Surface of Central Neurones

Z. Nusser and R. Shigemoto

10.1 Introduction

Most nerve cells in the central nervous system receive glutamate, the major excitatory neurotransmitter in the brain, from several distinct sources. For example, pyramidal cells in the CA1 area of the hippocampus receive a glutamatergic input from the entorhinal cortex in the stratum lacunosum-moleculare. These cells are also innervated by axon terminals of CA3 pyramidal cells, which also use glutamate as neurotransmitter, mainly in the strata oriens and radiatum, whereas local axon collaterals of the CA1 pyramidal cells are restricted to the stratum oriens and to the alveus (reviewed by Witter 1993). This multiple glutamatergic innervation of neurones is paralleled by the diversity of glutamate receptors (GluR) expressed by nerve cells. Three distinct classes of ionotropic GluRs [α-amino-3-hydroxy-5-methyl-4-isoxazolepropionate (AMPA)-type, kainate-type and *N*-methyl-D-aspartate (NMDA)-type] have been identified and classified according to their agonist selectivity, physiological properties and amino acid sequence homology between subunits constituting subtypes of these GluRs (reviewed by Seeburg 1993; Hollmann and Heinemann 1994). Several subunits have been discovered within each class, suggesting the existence of dozens of different ionotropic GluR subtypes. Furthermore, a large variety of metabotropic GluRs (mGluR) has also been revealed (reviewed by Nakanishi 1994; Pin and Duvoisin 1995). Pyramidal cells in the CA1 area express four AMPA-type GluR subunits (GluR1–4), two kainate-type GluR subunits (GluR6 and KA2), three subunits of the NMDA-type GluRs (NMDAR1, NMDAR2A and NMDAR2B) and two subtypes of the mGluRs (mGluR5 and mGluR7; Sato et al. 1993; Watanabe et al. 1993; Wisden and Seeburg 1993; Shigemoto et al. 1997).

The effect of excitatory inputs to nerve cells is finely balanced by inhibition. In the forebrain, the major neurotransmitter used for inhibition is γ-aminobutyric acid (GABA). The GABAergic innervation of most nerve cells and the molecular diversity of GABA receptors are at least as complex as those for glutamate. Pyramidal cells in the CA1 area of the hippocampus, our previous examples, are innervated by at least six distinct types of GABAergic interneurone (reviewed by Freund and Buzsaki 1996) and express nine subunits of ionotropic $GABA_A$ receptors (Persohn et al. 1992; Fritschy and Mohler 1995) and at least a $GABA_B$ receptor subtype (Kaupmann et al. 1997). Because of these multiple glutamatergic and GABAergic innervations of nerve cells and the expression of several distinct types of glutamate and GABA receptors, we were interested to establish whether every receptor subtype was distributed similarly on the surface of neurones. Or, receptor subtypes may be differentially targeted to distinct synapses, resulting in qualitative differences in the receptor content of synapses on individual cells. A third possibility is that different amounts of the same receptor may be present in distinct synapses of individual cells. To test these possibilities, we carried out immunogold localisation of several GABA and glutamate receptor subtypes at the electron microscopic level.

10.2 Differential Synaptic Targeting of the α1 and α2 Subunits of the $GABA_A$ Receptor on Hippocampal Pyramidal Cells

To determine whether every $GABA_A$ receptor subunit has a similar location on CA1 pyramidal cells or a selective distribution of subunits may occur, we have compared the precise subcellular location of the α1 and α2 subunits in relation to identified transmitter release sites (Nusser et al. 1996a). Immunopositive synapses for the α1 subunit were present on somata (Fig. 1A), proximal (Fig. 1B) and distal dendrites, spines (Fig. 1C) and axon initial segments (AIS; see Fig. 2D) of pyramidal cells. Quantitative evaluation revealed that approximately 65% and 55% of the synapses on somata/proximal dendrites and AISs, respectively, were immunopositive for the α1 subunit (Fig. 2A). The absence of immunolabelling in some of the symmetrical synapses on somata, dendrites and AISs may be due to either a lack of sensitivity of our method to reveal low concentrations of antigens or a genuine lack of this subunit from those synapses. The presence of this subunit in somatic and AIS synapses suggests that α1 subunit-containing $GABA_A$ receptors are activated by GABA released from both basket and axo-axonic cells. Furthermore, immunopositive dendritic synapses in strata oriens, radiatum and lacunosum-moleculare indicate that several dendrite targeting interneurones (reviewed by Freund and Buzsaki 1996) also act through α1 subunit-containing $GABA_A$ receptors. Although the majority of synapses on AISs were immunopositive for the α2 subunit (Fig. 1D,E), only a small proportion of the somatic and dendritic symmetrical synapses showed immunoreactivity for this subunit (Fig. 1F). Approximately 81% of the synapses on AISs contained the α2 subunit (Fig. 2B). The distribution of these synapses with regard to their α2 subunit content was not significantly different from normal (χ^2 test, p=0.134, n=52), suggesting that the immunonegative synapses on AISs

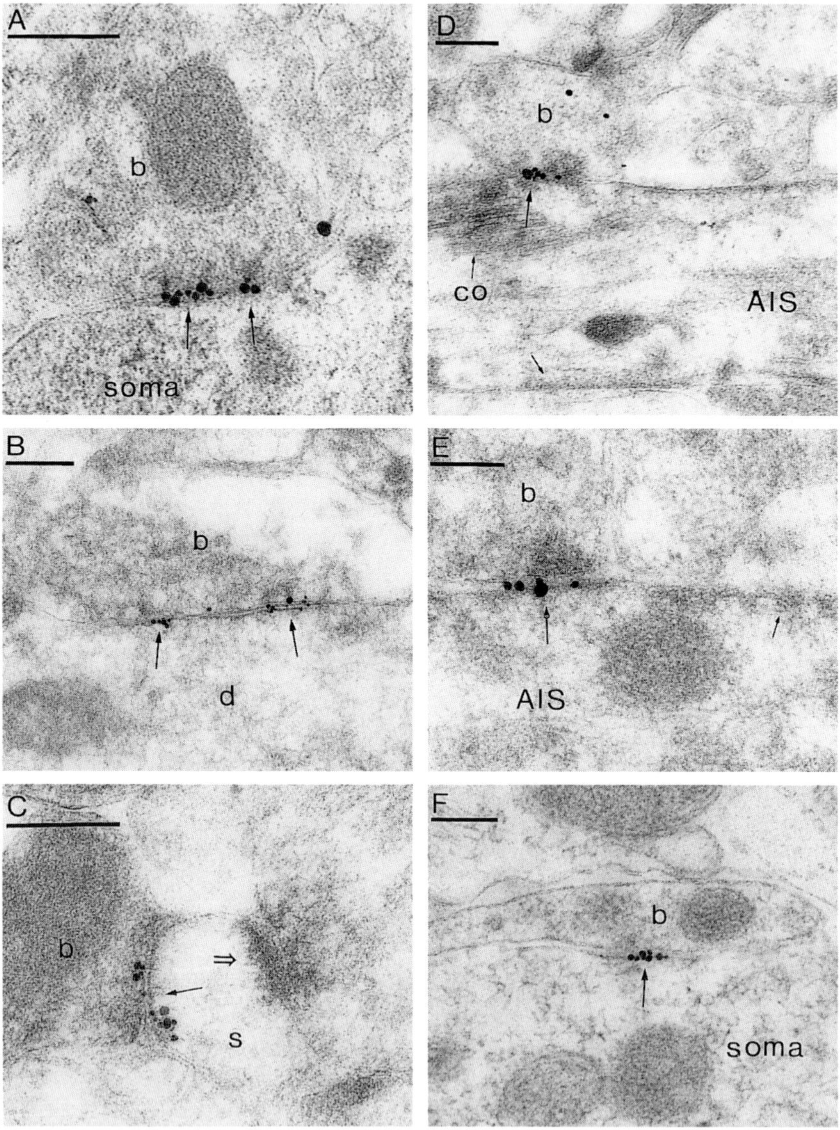

Fig. 1A–F. Legend see p. 191

do not form a separate population, but represent the lower tail of a single, immunopositive population. By contrast, there are two possible populations of synapses on somata/proximal dendrites with regard to their α2 subunit content. One of these, comprising approximately 17% of the synapses (Fig. 2B), contains a high density α2 subunit, whereas the other one (approximately 83% of synapses) has either no, or an undetectable level of the α2 subunit. These results demonstrate that, at least at the population level, these two α subunits are differentially distributed on pyramidal cells. To determine whether this differential synaptic location is also present in individual pyramidal cells, we serially sectioned the cell body layer of the CA1 area and immunoreacted one section for the α1 subunit and the consecutive one for the α2 subunit. In a representative cell shown in Fig. 2C, all three synapses on the AIS showed immunoreactivity for both studied subunits (Fig. 2C–E), but six out of seven synapses on the soma/basal dendrite contained immunoreactive α1 subunits only (Fig. 2C). Thus, the differential location of these two α subunits is also present on individual cells.

We have demonstrated that the α1 and α2 subunits are differentially targeted on the surface of hippocampal pyramidal cells. The α2 subunit is only present at detectable levels in a small subset of symmetrical synapses on somata and dendrites and in most synapses on AISs, whereas the α1 subunit is uniformly distributed in synapses over the axo-somato-dendritic domains. Kinetic and pharmacological properties of $GABA_A$ receptors are strongly influenced by the type of α subunit

◀ **Fig. 1. A–C** Electron micrographs showing immunoreactivity for the α1 subunit of the $GABA_A$ receptor in synapses (*arrows*) made by axon terminals (*b*) with a soma (**A**), a proximal dendrite (*d* in **B**) and a spine (*s* in **C**) of pyramidal cells in the CA1 area of the hippocampus. Post-embedding, silver-intensified immunogold reactions. Note an asymmetrical synapse (*open arrow*) on the spine in (**C**) is immunonegative for the $GABA_A$ receptor subunit. **D–F** Immunoreactivity for the α2 subunit of the $GABA_A$ receptor as demonstrated by post-embedding reactions on Lowicryl-embedded ultrathin sections. Axon terminals (*b*) establish immunopositive synaptic junctions (*arrows*) with axon initial segments (*AIS* in **D** and **E**), which can be recognised by their fine structural characteristics, e.g. membrane undercoating (*small arrows*), cisternal organelle (*co*). A rare α2 subunit immunopositive synapse (*arrow* in **F**) is shown between a bouton (*b*) and a pyramidal soma. Scales, 0.2 μm. (From Nusser et al. 1996a)

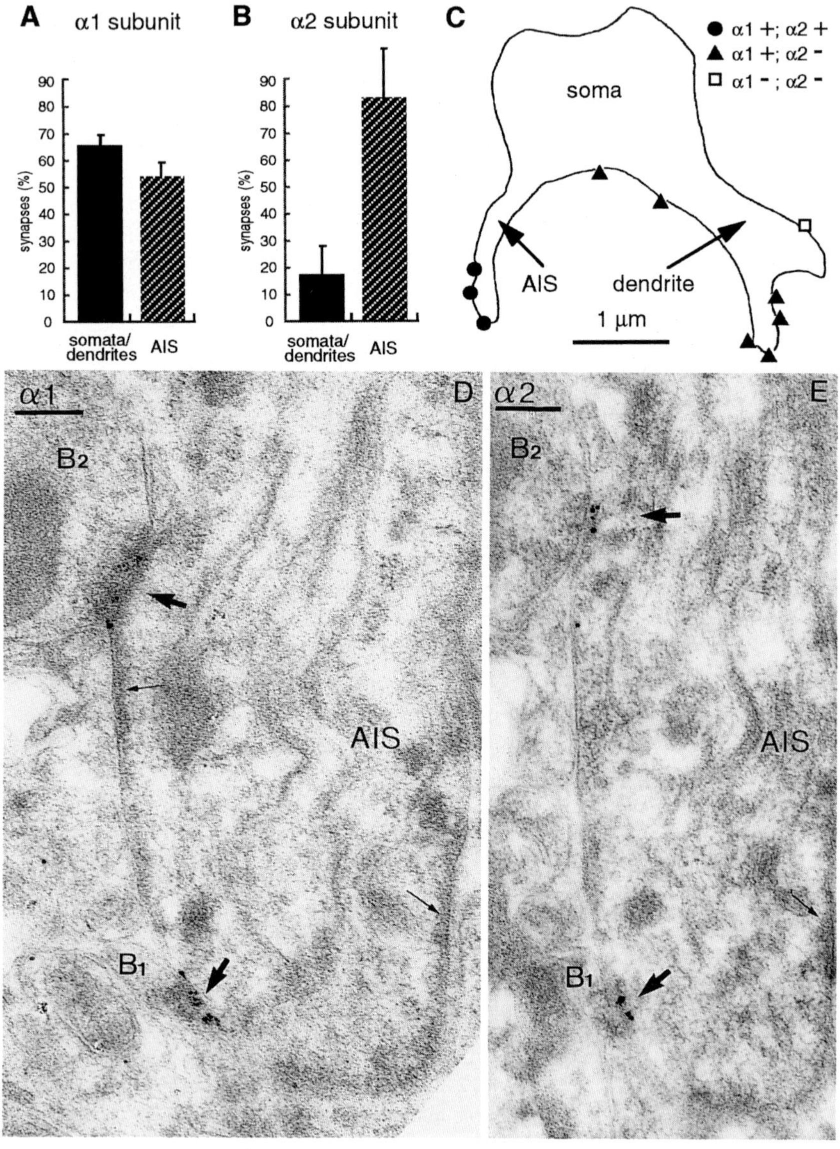

Fig. 2A–E. Legend see p. 193

incorporated (Pritchett et al. 1989; Saxena and Macdonald 1994; Luddens and Korpi 1995; Tia et al. 1996). Differences in the kinetics of either α1 or α2 subunit-containing $GABA_A$ receptors have not been revealed. However, when the α2 subunit is co-expressed with β and γ2 subunits, the receptors, in contrast to α1 subunit-containing ones, have a low affinity for several benzodiazepine ligands such as 2-oxoquazepam, zolpidem and the triazolopyridazine CL218872 (Pritchett et al. 1989; reviewed by Luddens and Wisden 1991; Sieghart 1995; Stephenson 1995). Thus, exogenous drugs and possible endogenous modulators of the benzodiazepine site are likely to affect the action of basket and axo-axonic cells differently. Differences in the role of these two types of GABAergic neurone in the network are as yet unknown, but the molecular difference revealed here between basket and axo-axonic cell synapses will help to establish the function of these cells.

◀ **Fig. 2. A,B** Quantitative distribution of synapses immunopositive for the α1 (**A**) and the α2 (**B**) subunits of the $GABA_A$ receptor on somata/proximal dendrites (*black columns*) or on axon initial segments (*AIS*; *striped columns*) of pyramidal cells in the CA1 area of rat hippocampus. Of the symmetrical synapses, 65.5% (4.2%) (mean, followed by standard deviation in parentheses) on somata/proximal dendrites (n=27) and 53.5% (5.0%) on AISs (n=15) are immunopositive for the α1 subunit. The majority, 81.4% (18.2%), of the synapses on AIS (n=52) are immunopositive for the α2 subunit, whereas only 17.1% (10.7%) of the symmetrical synapses (n=44) are immunopositive on somata/proximal dendrites. **C** Distribution of α1 and α2 subunits on an individual pyramidal cell. The AIS and one of the basal dendrites are emerging from the soma at a level where serial ultrathin sections were immunoreacted for both subunits. Three synapses on the AIS (*filled circles*) are immunopositive for both α subunits. Six out of seven symmetrical synapses on soma/basal dendrite are immunopositive for the α1 subunit alone (*triangles*) and only one synapse is immunonegative for both α subunits (*open square*). **D,E** Electron microscopic demonstration on serial sections of the co-localisation of the α1 and α2 subunits at synapses on the AIS shown in (**C**). Two boutons (B_1 and B_2) establish immunopositive synaptic junctions (*arrows*) for both the α1 (**D**) and α2 (**E**) subunits. *Small arrows* point to the undercoating of the plasma membrane, identifying the process as an AIS. Scales, 0.2 μm. (From Nusser et al. 1996a)

10.3 The α1 and α6 Subunits of the $GABA_A$ Receptor Have a Partially Differential Distribution on Cerebellar Granule Cells

Although cerebellar granule cells receive GABAergic input from Golgi cells only, they express six $GABA_A$ receptor subunits abundantly (α1, α6, β2, β3, γ2 and δ; Persohn et al. 1992; Wisden et al. 1992) which are co-assembled into at least three distinct $GABA_A$ receptors. We have applied the high-resolution immunogold localisation of the α1 and α6 subunits of the $GABA_A$ receptor to determine their precise location and relative density on granule cells (Nusser et al. 1996b). The α1 subunit was concentrated in GABAergic Golgi cell-to-granule cell synapses and was also present in the extrasynaptic dendritic and somatic membranes at a much lower density (see also Nusser et al. 1995). Asymmetrical synapses made by glutamatergic mossy fibre terminals with granule cell dendrites were always immunonegative for this subunit. The α6 subunit was also enriched in some of the GABAergic Golgi synapses and was found at a lower concentration in the extrasynaptic membranes. However, some of the excitatory mossy fibre-to-granule cell synapses showed an enrichment of immunoparticles for the α6 subunit which has never been observed for the α1 subunit. A cross-reactivity of our antibodies with unknown glutamate receptor-specific proteins is very unlikely (for controls see Nusser et al. 1996b) because: (a) Two different α6 subunit-specific antibodies gave identical results; (b) an enrichment of immunoparticles for the α6 subunit could not be observed in other excitatory synapses; (c) the labelling was not present in either Golgi or mossy synapses in α6-/- mice (Z. Nusser, W. Wisden and P. Somogyi, unpublished observation). Co-localisation experiments of the α1 and α6 subunits on consecutive ultrathin sections revealed that some of the Golgi synapses contained both α subunits, while others had only one or other of these subunits. In some mossy synapses, the enrichment of immunoparticles for the α6 subunit demonstrated that receptor immunoreactivity was preserved, but still no immunoreactivity for the α1 subunit could be detected in these synapses.

These results demonstrate a differential subcellular distribution of another two $GABA_A$ receptor α subunits in a central neurone. The functional significance of the presence of the α6 subunit in glutamatergic mossy synapses is as yet unknown, but several possibilities could be envisaged. Firstly, $\alpha6$ subunits in mossy synapses may not form functional GABA-gated Cl^- channels, therefore their presence has no functional consequence. Secondly, $\alpha6$ subunits may form functional $GABA_A$ receptors in combination with other types of subunits. These receptors could be activated by either GABA or other substances (e.g. β-alanine, γ-hydroxybutyrate, taurine) released from mossy fibre terminals. Since several studies have suggested that the overspill of glutamate to neighbouring synapses may play an important role in information processing (reviewed by Barbour and Hausser 1997), it is also possible that the $\alpha6$ subunit-containing receptors at mossy synapses are activated by GABA released at neighbouring Golgi synapses. Such a distant action of GABA would be similar to that observed in the hippocampus (Isaacson et al. 1993), where GABA released from GABAergic terminals acts on presynaptic $GABA_B$ receptors of glutamatergic axon terminals.

10.4 Differential Distribution of Immunoreactive $\alpha1$, $\alpha2$ and $\alpha3$ Subunits of the $GABA_A$ Receptor in Retinal α-Ganglion Cells

Koulen et al. (1996) have investigated the subcellular location of several subunits of the $GABA_A$ receptor on the surface of identified retinal α-ganglion cells to determine whether all expressed subunits had a similar distribution. α-ganglion cells were identified by intracellular injection of Lucifer Yellow and subsequently the retinae were immunostained for $GABA_A$ receptor subunits. Single α-ganglion cells contained intense immunoflourescence puncta for the $\alpha1$, $\alpha2$ and $\alpha3$ subunits in addition to the $\beta2/3$ and $\gamma2$ subunits. Furthermore, double-label immunofluorescence was applied to determine to what extent these $GABA_A$ receptor subunits were co-localised at postsynaptic sites. A similar pattern of labelling was observed when the $\alpha2$ subunit was co-localised with the $\gamma2$ subunit, indicating that some of the synapses contain both of these two subunits. However, when the sections were stained for the $\alpha1$ and $\alpha2$ subunits or $\alpha1$ and $\alpha3$ subunits, the average number of co-localisation of puncta was approximately 5%, which was not significantly larger than that expected from a random process. Therefore, the authors concluded that the $\alpha1$, $\alpha2$ and $\alpha3$ subunits were not co-localised at the same synapses. It is also suggested that synaptic clusters of $GABA_A$ receptors, containing different α subunits, might receive synaptic inputs from different types of amacrine cells. Because amacrine cells contain different neuroactive substances, a specific modulation of $GABA_A$ receptor function may occur at distinct sites.

10.5 Differential Localisation of δ and AMPA-Type GluRs in Cerebellar Purkinje Cells

Cerebellar Purkinje cells are innervated by two functionally distinct types of glutamatergic inputs, parallel and climbing fibres, and abundantly express δ, AMPA-type GluRs and mGluR1 (Hollmann et al. 1989; Masu et al. 1991; Yamazaki et al. 1992; Lomeli et al. 1993; Sato et al. 1993). In a fascinating study, Landsend et al. (1997) have demonstrated that more than 95% of immunogold particles for δ-type GluRs were associated with postsynaptic specialisations of cerebellar Purkinje cells and only a very low amount of these receptors could be detected at non-synaptic membranes. In addition, these receptors were selectively targeted to spines of Purkinje cells that were postsynaptic to parallel fibre terminals, but were absent from those spines which were postsynaptic to climbing-fibre terminals. By contrast, AMPA receptor subunits, GluR2/3, were present in both types of spines of Purkinje cells. These results demonstrate that neurones can compose distinct postsynaptic glutamate receptors, depending on the presynaptic input. How the δ GluR exerts its effect in cerebellar long-term depression, thus playing a role in motor co-ordination (Kashiwabuchi et al. 1995) is as yet unknown. However, this result suggests that it should be understood on the basis of a specific effect on parallel fibres to Purkinje cell synapses.

10.6 Target-Cell-Specific Concentration of mGluR7a in the Presynaptic Active Zone

The release of neurotransmitters from nerve terminals is regulated by presynaptic auto- or heteroreceptors. It has been shown that L-2-amino-4-phosphonobutyrate (AP4)-sensitive, type III mGluRs suppress the release of glutamate in the hippocampus (Gereau and Conn 1995; Manzoni and Bockaert 1995). Because only mGluR7a is extensively expressed in the hippocampus proper among type III mGluRs (Shigemoto et al. 1997), we have used a high-resolution immunogold localisation to determine its precise location in relation to transmitter release sites (Shigemoto et al. 1996). Some neuronal profiles were outlined by strong immunoreactive puncta, prominent over a weaker neuropile staining. Double immunolabelling experiments revealed that these neurones are mGluR1α-positive interneurones, which also express somatostatin and GABA (Baude et al. 1993). Electron microscopy revealed that immunogold particles for mGluR7a were restricted to the presynaptic specialisations of axon terminals, making asymmetrical (type I) synapses (Fig. 3a,b). Interestingly, a much higher density of immunoparticles was found on axon terminals making synapses with mGluR1α-immunopositive dendrites than on those synapsing with pyramidal cell spines or unidentified interneurone dendrites (Fig. 3b,c). The latter two groups contained either no, or very few, immunoparticles. There are two possible explanations for the above result. On one hand, mGluR1α-positive interneurones could be selectively innervated by a specific group of axons, making synapses only on these profiles and, therefore, every terminal would contain a high density of mGluR7a. Pyramidal cell spines would be innervated by another subset of axons which had a low level of mGluR7a content. On the other hand, it is also possible that a single axon may innervate both mGluR1α-immunopositive dendrites and pyramidal cell spines and would place a different amount of mGluR7a to its terminals, depending on the postsynaptic targets. To test the above possibilities, we identified single pyramidal cell axons by injecting *Phaseolus vulgaris* leucoagglutinin (PHAL) in the CA3 or CA1 area and labelled mGluR7a with immunoparticles. Single CA3 axons occasionally contacted mGluR7a decorated dendritic shafts and made several varicosities in the weakly labelled neuropile (Fig. 4a). We verified with electron microscopy that terminals of an identified pyramidal axon, making synapses with mGluR7a decorated dendrites, contained a much higher density of immunoparticles (Fig. 4c) than those of the same axon which contacted pyramidal cell spines (Fig. 4b,d). The most conspicuous example of the target-specific segregation of mGluR7a arose from the discovery that, when a single bouton made a release site to an mGluR7a-decorated dendrite, it contained a very high density of mGluR7a label, whereas another release site of the same

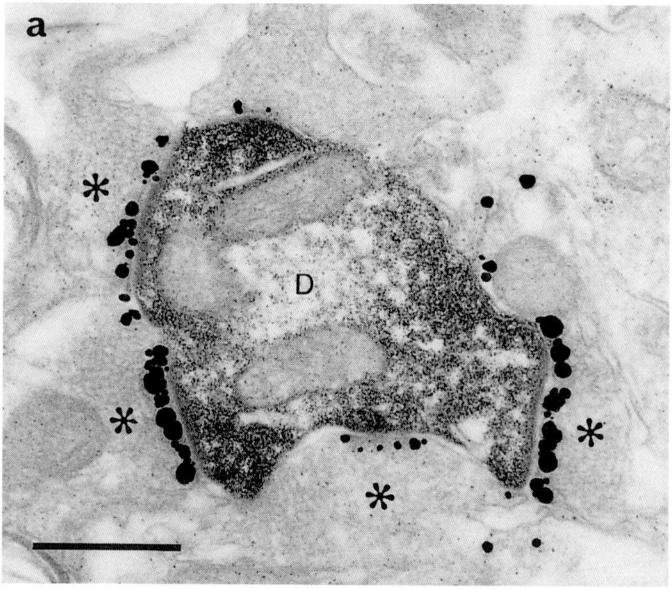

Fig. 3. a, b Electron microscopic demonstration of the high density of immunoparticles for mGluR7a at the presynaptic specialisations of synapses made by axon terminals (*asterisks*) with mGluR1α-immunopositive (peroxidase endproduct) dendrites (*D* and *D1*) in the CA3 area of rat hippocampus. Axon terminals making synapses with mGluR1α-negative dendrites (*D2* in **b**) or spines (not shown) are labelled only weakly, if at all. Scales, 0.4 μm. **c** Measurements of mGluR7a immunoreactivity show that axon terminals synapsing on mGluR1α immunopositive dendrites (*black columns*) have a much higher density of label than those making synapses on either pyramidal cell spines (*white columns*) or mGluR1α-negative dendrites (*grey column*). (From Shigemoto et al. 1996). **b,c** see p. 199

bouton synapsed with a pyramidal cell spine and had no label for mGluR7a (Fig. 5a).

It has been demonstrated that mGluR7a and other type-III mGluRs (Shigemoto et al. 1997) are restricted to presynaptic specialisations of axon terminals – the site of vesicle fusion. This distribution may reflect functional requirements such as being in precise conjunction with their target molecules, voltage-dependent Ca^{2+} channels (Takahashi et al. 1996), which are also thought to be concentrated at the presynaptic

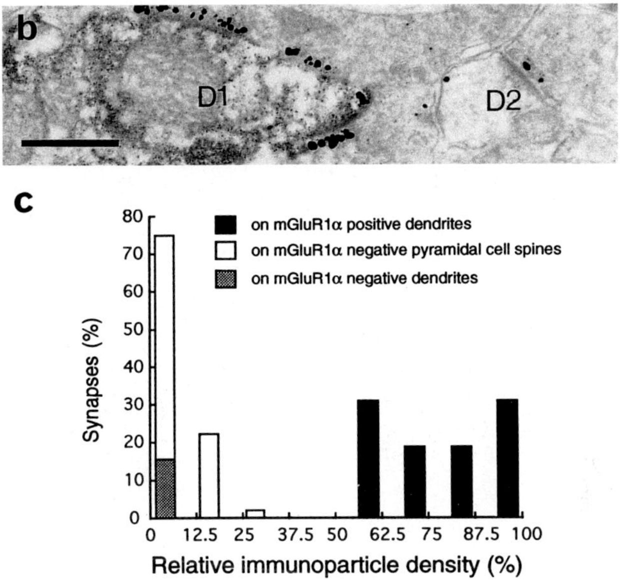

Fig. 3b,c. Legend see p. 198

active zone (Tareilus and Breer 1995). The apparently complete segregation of mGluR7a between two release sites of a single bouton (Fig. 5) suggests that coupling of the receptor to its effector is likely to be spatially restricted, and probably membrane delimited.

We have also demonstrated that the segregation of different amounts of mGluR7a can be present along individual axons and even within single boutons with multiple release sites and that the distinct levels of mGluR7a density depend on the identity of the postsynaptic elements. It seems that this target-specific segregation of presynaptic receptors, first described for mGluR7a (Shigemoto et al. 1996), may be a general principle of their organisation, because the same phenomenon has been reported (Shigemoto et al. 1997) for other mGluRs such as mGluR4a, mGluR7b and mGluR8. The probability of transmitter release may be different at synapses containing a high density of presynaptic mGluRs from what it is at those having either no or a very low density of mGluRs. If presynaptic mGluRs were persistently activated, it would

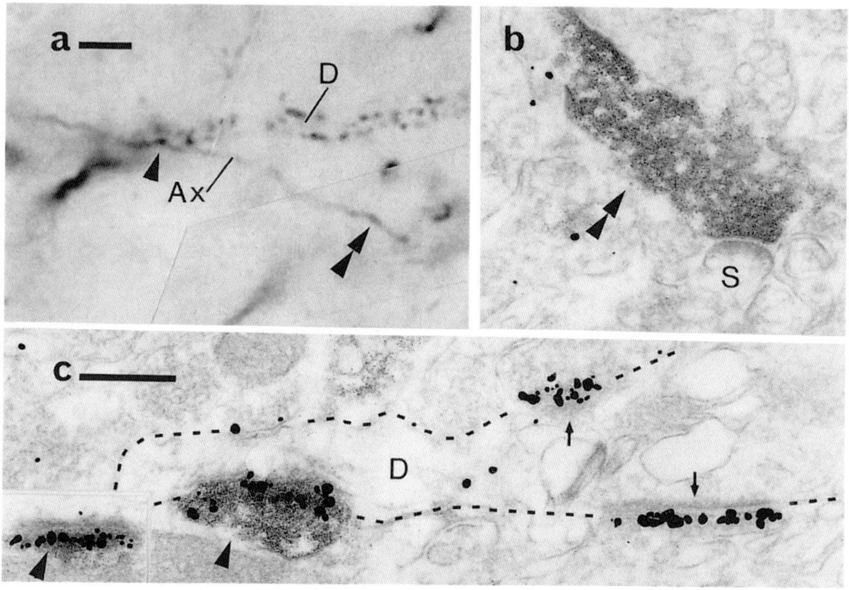

Fig. 4. a–c Corresponding light (**a**) and electron (**b, c**) micrographs showing that an identified CA3 pyramidal cell axon [anterograde labelling with *Phaseolus vulgaris* leucoagglutinin (PHAL); *Ax* in **a**; peroxidase product in **b** and **c**] makes synapses on a mGluR7a-decorated dendrite (*D* in **a** and **c**) and also (*double arrowheads* in **a**) on a pyramidal cell spine (*s* in **b**). One of the PHAL-labelled axon terminals (*arrowheads* in **a** and **c**) contains a high density of immunoparticles for mGluR7a (*inset* from a tilted consecutive section). Immunoparticles are not present in the other bouton of the same axon (*double arrowheads* in **a** and **b**) which make a synapse with a pyramidal cell spine. Scales: **a** 3 μm; **b, c** 0.3 μm. **d** Distinct levels of mGluR7a immunoreactivity in boutons of three identified pyramidal axons (*symbols*) according to the identity of postsynaptic targets. The level of mGluR7a in PHAL-labelled boutons is compared to that of other asymmetrical synapses around the identified boutons. Axon terminals on mGluR7a-decorated dendrites (*black columns*) contain a much higher density of labelling than those synapsing on unidentified dendrites (*grey column*) or on pyramidal cell spines (*white columns*). (From Shigemoto et al. 1996). **d** see p. 201

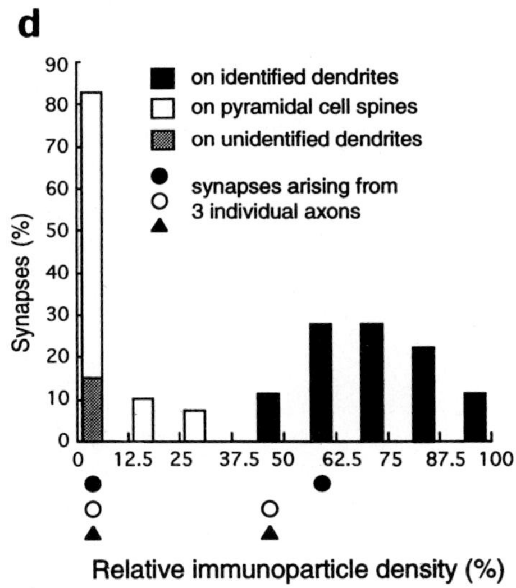

Fig. 4d. Legend see p. 200

result in a lower release probability at those synapses where a high density of presynaptic mGluR is present. Therefore, these synapses would be expected to transmit high frequency signals with a better efficiency than those with a low density of mGluRs. Indeed, in mGluR4-deficient mice, smaller responses were detected in a train of stimuli than those in wild-type animals (Pekhletski et al. 1996). In addition, it has been demonstrated that the effect of synapses on mGluR1α/somatostatin/GABA interneurones, where a high density of presynaptic mGluR7a is present, showed a much greater paired pulse and brief train facilitation than that of synapses on another type of interneurone of the hippocampus, where presynaptic mGluR density is low (Ali and Thomson 1997).

10.7 Conclusions

We have summarised recent results obtained with immunocytochemical localisation of glutamate and GABA receptors, demonstrating qualita-

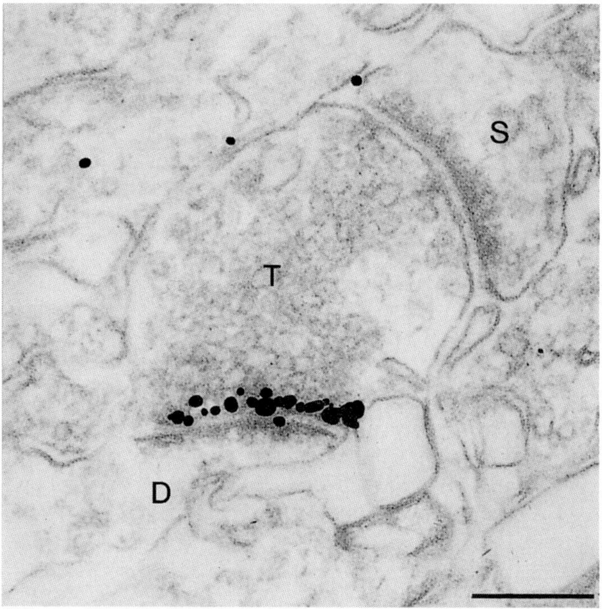

Fig. 5. The target-specific segregation of mGluR7a is demonstrated by an axon terminal (*T*) which makes two release sites. One of the release sites establishes a strongly receptor-immunopositive synapse with a dendritic shaft (*D*), characteristic of mGluR1α-positive neurones, whereas the other release site, which makes a synapse with a pyramidal cell spine (*S*), is immunonegative for mGluR7a. Scale, 0.2 μm. (From Shigemoto et al. 1996)

tive and quantitative differences in the receptor content of distinct synapses of individual nerve cells. Such differences are expected to result in functional heterogeneity of synapses according to the nature of pre- and postsynaptic structures. Therefore, the placement and precise origin of pre- and postsynaptic elements of synapses need to be determined for the interpretation of the mechanism of synaptic transmission and its modification with pharmacological agents.

Acknowledgements. This study was supported by the Medical Research Council (UK), a European Commission Shared Cost RTD Programme Grant (No. BIO4CT96-0585) and Japan Society for the Promotion of Science.

References

Ali AB, Thomson AM (1997) Brief train depression and facilitation at pyramidal-interneurone connections in slices of rat hippocampus; paired recordings with biocytin filling. J Physiol (Lond)501:9P

Barbour B, Hausser M (1997) Intersynaptic diffusion of neurotransmitter. Trends Neurosci 20:377–384

Baude A, Nusser Z, Roberts JDB, Mulvihill E, McIlhinney RAJ, Somogyi P (1993) The metabotropic glutamate receptor (mGluR1α) is concentrated at perisynaptic membrane of neuronal subpopulations as detected by immunogold reaction. Neuron 11:771–787

Freund TF, Buzsaki G (1996) Interneurons of the hippocampus. Hippocampus 6:347–470

Fritschy JM, Mohler H (1995) $GABA_A$-receptor heterogeneity in the adult rat brain: differential regional and cellular distribution of seven subunits. J Comp Neurol 359:154–194

Gereau RW, Conn PJ (1995) Multiple presynaptic metabotropic glutamate receptors modulate excitatory and inhibitory synaptic transmission in hippocampal area CA1. J Neurosci 15:6879–6889

Hollmann M, Heinemann S (1994) Cloned glutamate receptors. Annu Rev Neurosci 17:31–108

Hollmann M, O'Shea-Greenfield A, Rogers SW, Heinemann S (1989) Cloning by functional expression of a member of the glutamate receptor family. Nature 342:643–648

Isaacson JS, Solis JM, Nicoll RA (1993) Local and diffuse synaptic actions of GABA in the hippocampus. Neuron 10:165–175

Kashiwabuchi N, Ikeda K, Araki K, Hirano T, Shibuki K, Takayama C, Inoue Y, Kutsuwada T, Yagi T, Kang Y, Aizawa S, Mishina M (1995) Impairment of motor coordination, Purkinje cell synapse formation, and cerebellar long-term depression in GluRδ2 mutant mice. Cell 81:245–252

Kaupmann K, Huggel K, Heid J, Flor PJ, Bischoff S, Mickel SJ, McMaster G, Angst C, Bittiger H, Froestl W, Bettler B (1997) Expression cloning of $GABA_B$ receptors uncovers similarity to metabotropic glutamate receptors. Nature 386:239–246

Koulen P, Sassoe-Pognetto M, Grunert U, Wassle H (1996) Selective clustering of $GABA_A$ and glycine receptors in the mammalian retina. J Neurosci 16:2127–2140

Landsend AS, Amiry-Moghaddam M, Matsubara A, Bergersen L, Usami S, Whenthold RJ, Ottersen OP (1997) Differential localization of δ glutamate receptors in the rat cerebellum: coexpression with AMPA receptors in parallel fiber-spine synapses and absence from climbing fiber-spine synapses. J Neurosci 17:834–842

Lomeli H, Sprengel R, Laurie DJ, Kohr G, Herb A, Seeburg PH, Wisden W (1993) The rat delta-1 and delta-2 subunits extend the excitatory amino acid receptor family. FEBS Lett 315:318–322

Luddens H, Korpi ER (1995) GABA antagonists differentiate between recombinant $GABA_A$/benzodiazepine receptor subtypes. J Neurosci 15:6957–6962

Luddens H, Wisden W (1991) Function and pharmacology of multiple $GABA_A$ receptor subunits. Trends Pharmacol Sci 12:49–51

Manzoni O, Bockaert J (1995) Metabotropic glutamate receptors inhibiting excitatory synapses in the CA1 area of rat hippocampus. Eur J Neurosci 7:2518–2523

Masu M, Tanabe Y, Tsuchida K, Shigemoto R, Nakanishi S (1991) Sequence and expression of a metabotropic glutamate receptor. Nature 349:760–765

Nakanishi S (1994) Metabotropic glutamate receptors: synaptic transmission, modulation, and plasticity. Neuron 13:1031–1037

Nusser Z, Roberts JDB, Baude A, Richards JG, Somogyi P (1995) Relative densities of synaptic and extrasynaptic $GABA_A$ receptors on cerebellar granule cells as determined by a quantitative immunogold method. J Neurosci 15:2948–2960

Nusser Z, Sieghart W, Benke D, Fritschy J-M, Somogyi P (1996a) Differential synaptic localization of two major γ-aminobutyric acid type A receptor α subunits on hippocampal pyramidal cells. Proc Natl Acad Sci USA 93:11939–11944

Nusser Z, Sieghart W, Stephenson FA, Somogyi P (1996b) The $\alpha 6$ subunit of the $GABA_A$ receptor is concentrated in both inhibitory and excitatory synapses on cerebellar granule cells. J Neurosci 16:103–114

Pekhletski R, Gerlai R, Overstreet LS, Huang S, Agopyan N, Slater NT, Abramow-Newerly W, Roder JC, Hampson DR (1996) Impaired cerebellar synaptic plasticity and motor performance in mice lacking the mGluR4 subtype of metabotropic glutamate receptor. J Neurosci 16:6364–6373

Persohn E, Malherbe P, Richards JG (1992) Comparative molecular neuroanatomy of cloned $GABA_A$ receptor subunits in the rat CNS. J Comp Neurol 326:193–216

Pin JP, Duvoisin R (1995) The metabotropic glutamate receptors: structure and functions. Neuropharmacology 34:1–26

Pritchett DB, Luddens H, Seeburg PH (1989) Type I and type II $GABA_A$-benzodiazepine receptors produced in transfected cells. Science 245:1389–1392

Sato K, Kiyama H, Tohyama M (1993) The differential expression patterns of messenger RNAs encoding non-N-methil-D-aspartate receptor subunits (GluR1–4) in the rat brain. Neuroscience 52:515–539

Saxena NC, Macdonald RL (1994) Assembly of $GABA_A$ receptor subunits: role of the δ subunit. J Neurosci 14:7077–7086

Seeburg PH (1993) The molecular biology of mammalian glutamate receptor channels. Trends Neurosci 16:359–365

Shigemoto R, Kulik A, Roberts JDB, Ohishi H, Nusser Z, Kaneko T, Somogyi P (1996) Target-cell-specific concentration of a metabotropic glutamate receptor in the presynaptic active zone. Nature 381:523–525

Shigemoto R, Kinoshita A, Wada E, Nomura S, Ohishi H, Takada M, Flor PJ, Neki A, Abe T, Nakanishi S, Mizuno N (1997) Differential presynaptic localization of metabiotropic glutamate receptor subtypes in the rat hippocampus. J Neurosci 17:7503–7522

Sieghart W (1995) Structure and pharmacology of γ-aminobutyric acid$_A$ receptor subtypes. Pharmacol Rev 47:181–234

Stephenson FA (1995) The GABA$_A$ receptors. Biochem J 310:1–9

Takahashi T, Forsythe ID, Tsujimoto T, Barnes-Davies M, Onodera K (1996) Presynaptic calcium current modulation by a metabotropic glutamate receptor. Science 274:594–597

Tareilus E, Breer H (1995) Presynaptic calcium channels: pharmacology and regulation. Neurochem Int 26:539–558

Tia S, Wang JF, Kotchabhakdi N, Vicini S (1996) Distinct deactivation and desensitization kinetics of recombinant GABA$_A$ receptors. Neuropharmacology 35:1375–1382

Watanabe M, Inoue Y, Sakimura K, Mishina M (1993) Distinct distributions of five N-methyl-D-aspartate receptor channel subunit mRNAs in the forebrain. J Comp Neurol 338:377–390

Wisden W, Seeburg PH (1993) A complex mosaic of high-affinity kainate receptors in the brain. J Neurosci 13:3582–3598

Wisden W, Laurie DJ, Monyer H, Seeburg PH (1992) The distribution of 13 GABA$_A$ receptor subunit mRNAs in the rat brain. I. Telencephalon, diencephalon, mesencephalon. J Neurosci 12:1040–1062

Witter MP (1993) Organization of the entorhinal-hippocampal system: a review of current anatomical data. Hippocampus 3 [Special Issue]:33–44

Yamazaki M, Araki K, Shibata A, Mishina M (1992) Molecular cloning of a cDNA encoding a novel member of the mouse glutamate receptor channel family. Biochem Biophys Res Commun 183:886–892

Subject Index

Ernst Schering Research Foundation Workshop

Editors: Günter Stock
Ursula-F. Habenicht

Vol. 1 *(1991)*
Bioscience ⇆ Society
Workshop Report
Editors: D. J. Roy, B. E. Wynne, R. W. Old

Vol. 2 (1991)
Round Table Discussion on Bioscience ⇆ Society
Editor: J. J. Cherfas

Vol. 3 (1991)
Excitatory Amino Acids and Second Messenger Systems
Editors: V. I. Teichberg, L. Turski

Vol. 4 (1992)
Spermatogenesis – Fertilization – Contraception
Editors: E. Nieschlag, U.-F. Habenicht

Vol. 5 (1992)
Sex Steroids and the Cardiovascular System
Editors: P. Ramwell, G. Rubanyi, E. Schillinger

Vol. 6 (1993)
Transgenic Animals as Model Systems for Human Diseases
Editors: E. F. Wagner, F. Theuring

Vol. 7 (1993)
Basic Mechanisms Controlling Term and Preterm Birth
Editors: K. Chwalisz, R. E. Garfield

Vol. 8 (1994)
Health Care 2010
Editors: C. Bezold, K. Knabner

Vol. 9 (1994)
Sex Steroids and Bone
Editors: R. Ziegler, J. Pfeilschifter, M. Bräutigam

Vol. 10 (1994)
Nongenotoxic Carcinogenesis
Editors: A. Cockburn, L. Smith

Vol. 11 (1994)
Cell Culture in Pharmaceutical Research
Editors: N. E. Fusenig, H. Graf

Vol. 12 (1994)
Interactions Between Adjuvants, Agrochemical and Target Organisms
Editors: P. J. Holloway, R. T. Rees, D. Stock

Vol. 13 (1994)
Assessment of the Use of Single Cytochrome P450 Enzymes in Drug Research
Editors: M. R. Waterman, M. Hildebrand

Vol. 14 (1995)
Apoptosis in Hormone-Dependent Cancers
Editors: M. Tenniswood, H. Michna

Vol. 15 (1995)
Computer Aided Drug Design in Industrial Research
Editors: E. C. Herrmann, R. Franke

Vol. 16 (1995)
Organ-Selective Actions of Steroid Hormones
Editors: D. T. Baird, G. Schütz, R. Krattenmacher

Vol. 17 (1996)
Alzheimer's Disease
Editors: J.D. Turner, K. Beyreuther, F. Theuring

Vol. 18 (1997)
The Endometrium as a Target for Contraception
Editors: H.M. Beier, M.J.K. Harper, K. Chwalisz

Vol. 19 (1997)
EGF Receptor in Tumor Growth and Progression
Editors: R. B. Lichtner, R. N. Harkins

Vol. 20 (1997)
Cellular Therapy
Editors: H. Wekerle, H. Graf, J.D. Turner

Vol. 21 (1997)
Nitric Oxide, Cytochromes P 450, and Sexual Steroid Hormones
Editors: J.R. Lancaster, J.F. Parkinson

Vol. 22 (1997)
Impact of Molecular Biology and New Technical Developments in Diagnostic Imaging
Editors: W. Semmler, M. Schwaiger

Vol. 23 (1998)
Excitatory Amino Acids
Editors: P.H. Seeburg, I. Bresink, L. Turski

Supplement 1 (1994)
Molecular and Cellular Endocrinology of the Testis
Editors: G. Verhoeven, U.-F. Habenicht

Supplement 2 (1997)
Signal Transduction in Testicular Cells
Editors: V. Hansson, F. O. Levy, K. Taskén